# DEVRY INSTITUTES

# REVIEW GUIDE

# FOR

# MATHEMATICS

PRENTICE HALL, Upper Saddle River, NJ 07458

Editor: *Sally Denlow*
Production Editor: *Dawn Blayer*
Supplement Editor: *April Thrower*
Special Projects Manager: *Barbara A. Murray*
Production Coordinator: *Alan Fischer*
Cover Manager: *Paul Gourhan*

Printed in the United States of America

10  9  8  7  6  5

ISBN 0-13-095286-9

Prentice-Hall International (UK) Limited, *London*
Prentice-Hall of Australia Pty. Limited, *Sydney*
Prentice-Hall Canada, Inc., *Toronto*
Prentice-Hall Hispanoamericana, S.A., *Mexico*
Prentice-Hall of India Private Limited, *New Delhi*
Prentice-Hall of Japan, Inc., *Tokyo*
Prentice-Hall Asia Pte. Ltd., *Singapore*
Editora Prentice-Hall do Brasil, Ltda., *Rio de Janeiro*

# CONTENTS

# RATIONALS

## Decimals to Fractions

The centimeter is divided into tenths.
The length of line segment AB is 5 tenths,
or 0.5 centimeter.

$$0.5 = \frac{5}{10} \quad \begin{array}{l}\textbf{numerator} \\ \textbf{denominator}\end{array}$$

The **denominator** tells the number of equal parts in a whole.
The **numerator** tells the number of parts being considered.

The square is separated into 100 equal parts.
Six of the parts are shaded.

0.06 or $\frac{6}{100}$ of the square is shaded.

Ninety-four of the parts are not shaded.

0.94 or $\frac{94}{100}$ of the square is not shaded.

*Write the decimal and its equivalent fraction indicated by the shaded parts of each of the following.*

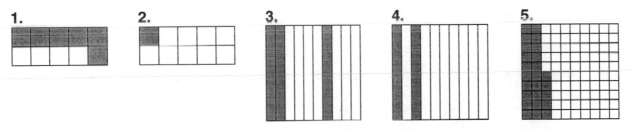

1.   2.   3.   4.   5.

*Write the decimal and its equivalent fraction or mixed numeral named below.*

6. 3 tenths      7. 20 hundredths    8. 34 hundredths    9. 308 thousandths

10. 7 tenths     11. 3 and 3 tenths   12. 60 hundredths 13. 9 hundredths

14. 2 and 6 hundredths    15. 2 thousandths    16. 2 hundredths    17. 10 tenths

18. 1 and 6 tenths     19. 605 thousandths 20. 81 hundredths   21. 19 thousandths

22. 3 and 857 thousandths 23. 7 and 3 tenths    24. 23 thousandths 25. 2 hundredths

*Copy and complete the following in words and give its equivalent fraction or mixed numeral.*

26. $32.47 = 32\frac{47}{100}$    27. 0.9 = nine tenths   28. 0.83    29. 6.7     30. 0.171

31. 1.075        32. 5.06           33. 0.541   34. 12.09   35. 0.001

1

# Greatest Common Factor

A figure with an area of 16 square units can be shown in several ways.

**16 ÷ 1 = 16**

**16 ÷ 2 = 8**

**16 ÷ 4 = 4**

The factors of 16 are the numbers by which 16 can be divided evenly. The factors of 16 are 1, 2, 4, 8, and 16.

What are the factors of 14?

**14 ÷ 1 = 14**

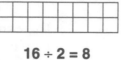

**14 ÷ 2 = 7**

The factors of 14 are 1, _____, _____, and _____.

Find the greatest common factor (GCF) of 16 and 14.

**Factors of 16** $^y$    **1**    **2**    4    8    16

**Factors of 14** $^y$    **1**    **2**    7    14

The shaded numbers are their common factors.

What is the greatest common factor? _____

*List the factors of each of the following numbers.*

**1.** 9      **2.** 8      **3.** 12      **4.** 15      **5.** 30      **6.** 18

*Find the GCF for each pair of numbers.*

**7.** 8 and 10      **8.** 2 and 10      **9.** 9 and 12      **10.** 5 and 10

**11.** 3 and 7      **12.** 6 and 18      **13.** 5 and 25      **14.** 4 and 12

*Find the GCF of the numerator and denominator of each fraction below.*

**15.** $\frac{9}{12}$      **16.** $\frac{6}{9}$      **17.** $\frac{9}{15}$      **18.** $\frac{14}{16}$      **19.** $\frac{12}{18}$

**20.** $\frac{10}{12}$      **21.** $\frac{15}{20}$      **22.** $\frac{3}{18}$      **23.** $\frac{10}{30}$      **24.** $\frac{4}{38}$

**25.** $\frac{25}{55}$      **26.** $\frac{9}{48}$      **27.** $\frac{16}{24}$      **28.** $\frac{40}{50}$      **29.** $\frac{63}{72}$

**30.** $\frac{42}{66}$      **31.** $\frac{39}{51}$      **32.** $\frac{45}{105}$      **33.** $\frac{26}{72}$      **34.** $\frac{84}{90}$

# Simplifying Fractions

Change $\frac{40}{56}$ to simplest form.

Find the GCF of 40 and 56.

List the factors.

40 $^y$  1, 2, 4, 5, **8**, 10, 20, 40

56 $^y$  1, 2, 4, 7, **8**, 14, 28, 56

The GCF is 8.

Divide both 40 and 56 by 8.

$$\frac{40}{56} \quad \begin{array}{c} \div 8 \\ \\ \frac{5}{7} \end{array} \div 8$$

Can you find any number other than 1 that divides both 5 and 7 evenly? _____

Divide both the numerator and denominator by their GCF to change a fraction to simplest form.

> **A fraction is in simplest form when the greatest common factor (GCF) of its numerator and denominator is 1.**

*Simplify each fraction not in simplest form.*

1. $\frac{5}{10}$  2. $\frac{2}{6}$  3. $\frac{4}{12}$  4. $\frac{6}{24}$  5. $\frac{6}{10}$

6. $\frac{6}{18}$  7. $\frac{10}{15}$  8. $\frac{9}{24}$  9. $\frac{18}{24}$  10. $\frac{14}{18}$

11. $\frac{6}{9}$  12. $\frac{9}{12}$  13. $\frac{3}{5}$  14. $\frac{12}{16}$  15. $\frac{8}{12}$

16. $\frac{10}{25}$  17. $\frac{12}{14}$  18. $\frac{15}{18}$  19. $\frac{10}{16}$  20. $\frac{6}{16}$

21. $\frac{30}{36}$  22. $\frac{4}{18}$  23. $\frac{18}{45}$  24. $\frac{21}{49}$  25. $\frac{3}{10}$

26. $\frac{56}{64}$  27. $\frac{3}{21}$  28. $\frac{26}{39}$  29. $\frac{35}{42}$  30. $\frac{14}{35}$

31. $\frac{24}{40}$  32. $\frac{22}{99}$  33. $\frac{7}{11}$  34. $\frac{7}{35}$  35. $\frac{36}{54}$

36. $\frac{24}{42}$  37. $\frac{3}{12}$  38. $\frac{80}{90}$  39. $\frac{45}{60}$  40. $\frac{45}{80}$

# Improper Fractions

The improper fraction $\frac{8}{5}$ means $8 \div 5$.

$$\frac{8}{5} \;\rightarrow\; 5\overline{)\,\begin{array}{r}1\phantom{0}\\8\\-5\\\hline 3\end{array}} \;\rightarrow\; 1\frac{3}{5}$$

Write the remainder, 3, as the fraction $\frac{3}{5}$.
The denominator, 5, is the divisor.
Written as a mixed numeral, $\frac{8}{5}$ is $1\frac{3}{5}$.

*Write a division problem for each of the following.*

1. $\frac{9}{4}$  2. $\frac{8}{3}$  3. $\frac{17}{5}$  4. $\frac{16}{10}$

5. $\frac{19}{9}$  6. $\frac{7}{4}$  7. $\frac{36}{3}$  8. $\frac{169}{13}$

9. $\frac{60}{3}$  10. $\frac{12}{4}$  11. $\frac{59}{13}$  12. $\frac{131}{19}$

*Write each improper fraction as a mixed numeral or whole number in simplest form.*

13. $\frac{9}{4}$  14. $\frac{9}{8}$  15. $\frac{13}{9}$  16. $\frac{4}{3}$

17. $\frac{7}{3}$  18. $\frac{24}{3}$  19. $\frac{15}{7}$  20. $\frac{45}{9}$

21. $\frac{7}{6}$  22. $\frac{10}{3}$  23. $\frac{19}{17}$  24. $\frac{22}{7}$

25. $\frac{8}{5}$  26. $\frac{10}{6}$  27. $\frac{15}{5}$  28. $\frac{16}{12}$

29. $\frac{6}{4}$  30. $\frac{15}{6}$  31. $\frac{11}{4}$  32. $\frac{15}{10}$

33. $\frac{72}{8}$  34. $\frac{10}{8}$  35. $\frac{56}{7}$  36. $\frac{14}{4}$

37. $\frac{10}{6}$  38. $\frac{12}{10}$  39. $\frac{16}{12}$  40. $\frac{14}{10}$

41. $\frac{33}{9}$  42. $\frac{18}{4}$  43. $\frac{40}{16}$  44. $\frac{23}{12}$

45. $\frac{58}{8}$  46. $\frac{32}{12}$  47. $\frac{47}{6}$  48. $\frac{32}{6}$

49. $\frac{12}{5}$  50. $\frac{3}{2}$  51. $\frac{5}{4}$  52. $\frac{9}{6}$

53. $\frac{9}{7}$  54. $\frac{9}{2}$  55. $\frac{11}{4}$  56. $\frac{14}{10}$

57. $\frac{6}{4}$  58. $\frac{55}{20}$  59. $\frac{55}{15}$  60. $\frac{40}{16}$

# Addition and Subtraction of Fractions

To add or subtract fractions which have like denominators, add or subtract the numerators. Then, write the sum or difference over the denominator.

Add $\frac{5}{9} + \frac{2}{9}$.

$$\frac{5}{9} + \frac{2}{9} = \frac{5+2}{9} = \frac{7}{9}$$

Subtract $\frac{1}{4}$ from $\frac{3}{4}$.

$$\frac{3}{4} - \frac{1}{4} = \frac{3-1}{4} = \frac{2}{4} = \frac{1}{2}$$

*Add or subtract. Write the sum or difference in simplest form.*

1. $\frac{1}{5} + \frac{4}{5}$

2. $\frac{3}{8} + \frac{7}{8}$

3. $\frac{9}{10} + \frac{1}{10}$

4. $\frac{4}{7} + \frac{3}{7}$

5. $\frac{7}{9} - \frac{4}{9}$

6. $\frac{13}{15} - \frac{9}{15}$

7. $\frac{3}{7} - \frac{1}{7}$

8. $\frac{10}{11} - \frac{9}{11}$

9. $\frac{3}{4} + \frac{1}{4}$

10. $\frac{4}{6} + \frac{2}{6}$

11. $\frac{7}{8} + \frac{2}{8}$

12. $\frac{4}{9} + \frac{4}{9}$

13. $\frac{6}{18} - \frac{3}{18}$

14. $\frac{8}{16} - \frac{3}{16}$

15. $\frac{15}{30} - \frac{8}{30}$

16. $\frac{15}{35} - \frac{13}{35}$

17. $\frac{9}{20} + \frac{3}{20}$

18. $\frac{4}{18} + \frac{14}{18}$

19. $\frac{4}{10} + \frac{5}{10}$

20. $\frac{3}{8} + \frac{5}{8}$

21. $\frac{17}{18} - \frac{5}{18}$

22. $\frac{4}{7} + \frac{6}{7}$

23. $\frac{20}{21} - \frac{8}{21}$

24. $\frac{5}{9} + \frac{8}{9}$

25. $\frac{29}{36} - \frac{11}{36}$

26. $\frac{2}{3} + \frac{2}{3}$

27. $\frac{15}{32} - \frac{10}{32}$

28. $\frac{4}{15} - \frac{1}{15}$

29. $\frac{5}{11} - \frac{1}{11}$

30. $\frac{3}{4} + \frac{3}{4}$

31. $\frac{8}{15} + \frac{11}{15}$

32. $\frac{7}{10} - \frac{1}{10}$

33. $\frac{23}{28} - \frac{9}{28}$

34. $\frac{11}{24} + \frac{23}{24}$

35. $\frac{8}{9} - \frac{2}{9}$

36. $\frac{10}{21} + \frac{17}{21}$

37. $\frac{5}{6} + \frac{5}{6}$

38. $\frac{1}{2} - \frac{1}{2}$

39. $\frac{27}{30} - \frac{12}{30}$

40. $\frac{16}{20} - \frac{12}{20}$

41. $\frac{21}{24} + \frac{9}{24}$

42. $\frac{15}{25} + \frac{20}{25}$

43. $\frac{8}{12} + \frac{10}{12}$

44. $\frac{10}{16} + \frac{14}{16}$

# Equivalent Fractions

The matching figures below are shaded the same but have a different number of equivalent parts.

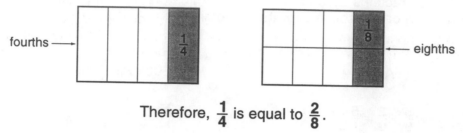

fourths ⟶

eighths ⟵

Therefore, $\frac{1}{4}$ is equal to $\frac{2}{8}$.

*Find equivalent fractions for the amounts shaded.*

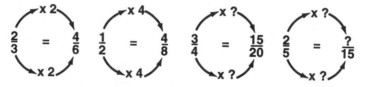

1. $\frac{\square}{2} = \frac{\square}{8}$

2. $\frac{2}{3} = \frac{\square}{6}$

3. $\frac{\square}{8} = \frac{\square}{16}$

Multiply both the numerator and denominator by the same nonzero number to find equivalent fractions.

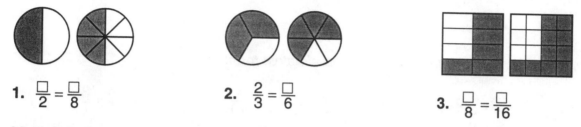

$\frac{2}{3} = \frac{4}{6}$ (×2) $\qquad$ $\frac{1}{2} = \frac{4}{8}$ (×4) $\qquad$ $\frac{3}{4} = \frac{15}{20}$ (×?) $\qquad$ $\frac{2}{5} = \frac{?}{15}$ (×?)

*Find the number used to multiply the numerator and denominator so that the fractions are equivalent.*

4. $\frac{1}{3} = \frac{4}{12}$

5. $\frac{3}{5} = \frac{9}{15}$

6. $\frac{5}{7} = \frac{25}{35}$

7. $\frac{5}{6} = \frac{15}{18}$

8. $\frac{7}{10} = \frac{14}{20}$

9. $\frac{1}{4} = \frac{4}{16}$

10. $\frac{2}{5} = \frac{24}{60}$

11. $\frac{2}{9} = \frac{16}{72}$

*Complete to make the fractions equivalent.*

12. $\frac{4}{5} = \frac{\square}{25}$

13. $\frac{5}{8} = \frac{\square}{24}$

14. $\frac{3}{4} = \frac{\square}{16}$

15. $\frac{9}{10} = \frac{\square}{30}$

16. $\frac{10}{12} = \frac{40}{\square}$

17. $\frac{2}{14} = \frac{8}{\square}$

18. $\frac{2}{7} = \frac{14}{\square}$

19. $\frac{3}{16} = \frac{27}{\square}$

20. $\frac{5}{7} = \frac{\square}{42}$

21. $\frac{6}{9} = \frac{12}{\square}$

22. $\frac{1}{13} = \frac{\square}{39}$

23. $\frac{4}{13} = \frac{44}{\square}$

24. $\frac{1}{2} = \frac{8}{\square}$

25. $\frac{5}{6} = \frac{\square}{12}$

26. $\frac{2}{3} = \frac{8}{\square}$

27. $\frac{2}{7} = \frac{\square}{14}$

28. $\frac{1}{3} = \frac{\square}{27}$

29. $\frac{5}{8} = \frac{15}{\square}$

30. $\frac{2}{3} = \frac{\square}{9}$

31. $\frac{3}{5} = \frac{\square}{20}$

32. $\frac{3}{8} = \frac{12}{\square}$

33. $\frac{9}{10} = \frac{45}{\square}$

34. $\frac{2}{5} = \frac{\square}{30}$

35. $\frac{1}{4} = \frac{\square}{52}$

# Least Common Multiple

Which names the greater number, $\frac{3}{8}$ or $\frac{5}{12}$?

Change the fractions to equivalent fractions with the same denominator.
Then, compare the numerators. In order to do this, you must find the least common multiple (LCM) of the two denominators. There are two ways to find the LCM.

Method A (using multiples)

Multiples of 8 are 8, 16, 24, 32, 40, 48 ...
Multiples of 12 are 12, 24, 36, 48, 60 ...
Common multiples are 24 and 48.
The least common multiple (LCM) is 24.

Change to equivalent fractions that
have the LCM as the denominator.
Compare the numerators to determine
which factor names the greater numbers.

$\frac{3}{8} = \frac{9}{24}$ $\qquad$ $\frac{5}{12} = \frac{10}{24}$

Since $\frac{9}{24} < \frac{10}{24}$, then $\frac{3}{8} < \frac{5}{12}$.

*Find the first 5 multiples of each number.*

1. 6

2. 5

3. 4

4. 3

5. 15

6. 9

7. 14

8. 10

9. 7

*Find the LCM for the denominators of each pair of fractions.*

10. $\frac{2}{3}, \frac{1}{2}$ _____

11. $\frac{1}{3}, \frac{5}{6}$ _____

12. $\frac{1}{6}, \frac{5}{8}$ _____

13. $\frac{1}{4}, \frac{1}{6}$ _____

14. $\frac{3}{5}, \frac{5}{8}$ _____

15. $\frac{5}{6}, \frac{1}{15}$ _____

16. $\frac{4}{7}, \frac{4}{9}$ _____

17. $\frac{1}{6}, \frac{9}{10}$ _____

18. $\frac{1}{8}, \frac{3}{18}$ _____

19. $\frac{2}{7}, \frac{7}{12}$ _____

*Use the LCM to rename each pair of fractions. Then use <, >, or = to compare them.*

20. $\frac{5}{7}, \frac{3}{4}$

21. $\frac{3}{8}, \frac{4}{9}$

22. $\frac{3}{14}, \frac{1}{6}$

23. $\frac{4}{5}, \frac{3}{4}$

24. $\frac{1}{2}, \frac{3}{5}$

25. $\frac{10}{12}, \frac{7}{8}$

26. $\frac{7}{20}, \frac{6}{15}$

27. $\frac{8}{9}, \frac{5}{6}$

28. $\frac{9}{10}, \frac{15}{18}$

29. $\frac{2}{3}, \frac{3}{4}$

A second method for finding the LCM involves prime numbers and composite numbers.

A **prime number** has exactly
two factors, one and itself.

$17 = 1 \times 17$
The only factors of 17
are 1 and 17.

17 is a prime number.

A **composite number** has
more than two factors.

$16 = 1 \times 16$, $16 = 2 \times 8$
$16 = 4 \times 4$
The factors of 16 are
1, 2, 4, 8, and 16.

16 is a composite
number.

State whether each number is prime or composite.

**30.** 32      **31.** 16      **32.** 41      **33.** 64      **34.** 51      **35.** 35

**36.** 12      **37.** 6      **38.** 20      **39.** 29      **40.** 47      **41.** 39

Complete each factor tree to find the prime factors of each number.

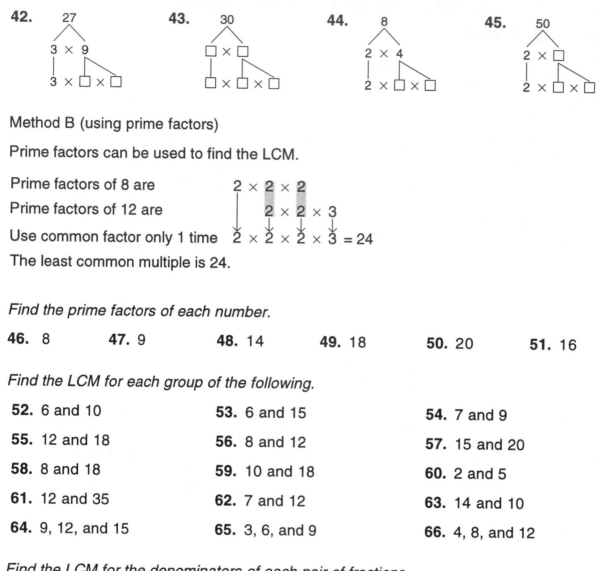

Method B (using prime factors)

Prime factors can be used to find the LCM.

Prime factors of 8 are      $2 \times 2 \times 2$

Prime factors of 12 are      $2 \times 2 \times 3$

Use common factor only 1 time   $2 \times 2 \times 2 \times 3 = 24$

The least common multiple is 24.

Find the prime factors of each number.

**46.** 8      **47.** 9      **48.** 14      **49.** 18      **50.** 20      **51.** 16

Find the LCM for each group of the following.

**52.** 6 and 10      **53.** 6 and 15      **54.** 7 and 9

**55.** 12 and 18      **56.** 8 and 12      **57.** 15 and 20

**58.** 8 and 18      **59.** 10 and 18      **60.** 2 and 5

**61.** 12 and 35      **62.** 7 and 12      **63.** 14 and 10

**64.** 9, 12, and 15      **65.** 3, 6, and 9      **66.** 4, 8, and 12

Find the LCM for the denominators of each pair of fractions.

**67.** $\frac{2}{3}, \frac{4}{7}$    **68.** $\frac{3}{4}, \frac{2}{5}$    **69.** $\frac{2}{9}, \frac{1}{5}$    **70.** $\frac{5}{12}, \frac{7}{8}$    **71.** $\frac{3}{8}, \frac{1}{6}$

**72.** $\frac{1}{4}, \frac{3}{7}$    **73.** $\frac{5}{9}, \frac{3}{4}$    **74.** $\frac{3}{12}, \frac{3}{10}$    **75.** $\frac{3}{20}, \frac{2}{15}$    **76.** $\frac{4}{15}, \frac{1}{6}$

Use the LCM to rename each pair of fractions. Then use <, >, or = to compare them.

**77.** $\frac{2}{5}, \frac{7}{10}$    **78.** $\frac{3}{10}, \frac{6}{20}$    **79.** $\frac{2}{3}, \frac{3}{7}$    **80.** $\frac{3}{4}, \frac{5}{9}$    **81.** $\frac{3}{14}, \frac{7}{20}$

# Addition and Subtraction of Fractions

To add or subtract fractions with unlike denominators, first rename the fractions with like denominators. Then, add or subtract.

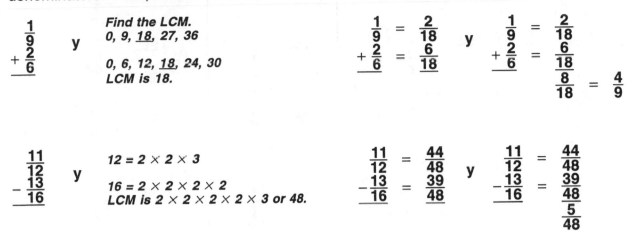

$\frac{1}{9}$
$+\frac{2}{6}$

y

**Find the LCM.**
0, 9, *18*, 27, 36

0, 6, 12, *18*, 24, 30
**LCM is 18.**

$\frac{1}{9} = \frac{2}{18}$
$+\frac{2}{6} = \frac{6}{18}$

y

$\frac{1}{9} = \frac{2}{18}$
$+\frac{2}{6} = \frac{6}{18}$
$\frac{8}{18} = \frac{4}{9}$

$\frac{11}{12}$
$-\frac{13}{16}$

y

$12 = 2 \times 2 \times 3$

$16 = 2 \times 2 \times 2 \times 2$
**LCM is $2 \times 2 \times 2 \times 2 \times 3$ or 48.**

$\frac{11}{12} = \frac{44}{48}$
$-\frac{13}{16} = \frac{39}{48}$

y

$\frac{11}{12} = \frac{44}{48}$
$-\frac{13}{16} = \frac{39}{48}$
$\frac{5}{48}$

*Add or subtract. Write the sum or difference in simplest form.*

1. $\frac{1}{4}$
   $-\frac{1}{5}$

2. $\frac{1}{2}$
   $-\frac{1}{3}$

3. $\frac{3}{5}$
   $-\frac{1}{3}$

4. $\frac{7}{8}$
   $-\frac{1}{6}$

5. $\frac{1}{4}$
   $+\frac{1}{8}$

6. $\frac{1}{3}$
   $+\frac{5}{6}$

7. $\frac{2}{9}$
   $+\frac{2}{3}$

8. $\frac{3}{4}$
   $+\frac{5}{8}$

9. $\frac{3}{4}$
   $-\frac{2}{3}$

10. $\frac{2}{3}$
    $-\frac{1}{4}$

11. $\frac{6}{7}$
    $-\frac{2}{3}$

12. $\frac{5}{6}$
    $-\frac{2}{9}$

13. $\frac{1}{8}$
    $+\frac{5}{6}$

14. $\frac{2}{5}$
    $+\frac{3}{7}$

15. $\frac{3}{4}$
    $+\frac{2}{5}$

16. $\frac{1}{3}$
    $+\frac{3}{8}$

*Solve each problem.*

17. Oklahoma City has $\frac{3}{20}$ of the population of Oklahoma. Tulsa has $\frac{3}{25}$ of the population of Oklahoma. What part of the population of Oklahoma do these cities have?

18. Phoenix has $\frac{3}{10}$ of the population of Arizona. Tucson has $\frac{3}{20}$ of the population of Arizona. What part of the population of Arizona do these cities have?

Add or subtract. Write the sum or difference in simplest form.

19. $\dfrac{5}{6}$
$+\dfrac{1}{4}$

20. $\dfrac{3}{8}$
$+\dfrac{5}{6}$

21. $\dfrac{5}{6}$
$+\dfrac{2}{9}$

22. $\dfrac{6}{7}$
$-\dfrac{1}{3}$

23. $\dfrac{7}{9}$
$-\dfrac{1}{2}$

24. $\dfrac{3}{7}$
$-\dfrac{2}{5}$

25. $\dfrac{8}{15}$
$+\dfrac{1}{5}$

26. $\dfrac{7}{8}$
$-\dfrac{5}{6}$

27. $\dfrac{1}{5}$
$+\dfrac{5}{6}$

28. $\dfrac{5}{8}$
$+\dfrac{5}{6}$

29. $\dfrac{11}{12}$
$+\dfrac{2}{3}$

30. $\dfrac{6}{7}$
$-\dfrac{1}{3}$

31. $\dfrac{13}{24}$
$-\dfrac{1}{8}$

32. $\dfrac{3}{7}$
$-\dfrac{5}{21}$

33. $\dfrac{14}{15}$
$-\dfrac{1}{3}$

34. $\dfrac{11}{12}$
$-\dfrac{2}{3}$

35. $\dfrac{2}{3}$
$+\dfrac{3}{7}$

36. $\dfrac{7}{9}$
$+\dfrac{3}{8}$

37. $\dfrac{7}{20}$
$+\dfrac{4}{5}$

38. $\dfrac{4}{5}$
$-\dfrac{11}{20}$

39. $\dfrac{9}{10}$
$-\dfrac{2}{9}$

40. $\dfrac{11}{12}$
$-\dfrac{2}{5}$

41. $\dfrac{3}{7}$
$+\dfrac{2}{5}$

42. $\dfrac{2}{9}$
$+\dfrac{1}{6}$

43. $\dfrac{5}{9}$
$+\dfrac{2}{5}$

44. $\dfrac{3}{4}$
$+\dfrac{3}{10}$

45. $\dfrac{6}{7}$
$-\dfrac{2}{9}$

46. $\dfrac{14}{15}$
$-\dfrac{5}{6}$

47. $\dfrac{5}{6}$
$-\dfrac{1}{4}$

48. $\dfrac{7}{8}$
$-\dfrac{7}{12}$

49. $\dfrac{9}{10}$
$-\dfrac{5}{12}$

50. $\dfrac{5}{6}$
$-\dfrac{3}{8}$

51. $\dfrac{1}{3}$
$+\dfrac{3}{5}$

52. $\dfrac{5}{6}$
$+\dfrac{2}{5}$

53. $\dfrac{5}{9}$
$+\dfrac{7}{8}$

54. $\dfrac{1}{3}$
$+\dfrac{3}{7}$

Solve each problem.

55. Trucks use $\dfrac{1}{4}$ of the transportation energy. Airplanes use $\dfrac{1}{20}$. What part of the transportation energy is used by trucks and airplanes?

56. Cars use $\dfrac{1}{2}$ of the transportation energy. Buses and railroads use $\dfrac{1}{6}$. What part of the transportation energy is used by cars, buses, and railroads?

# Addition and Subtraction of Mixed Numerals

To add or subtract mixed numerals with unlike denominators, first rename the fractions with like denominators. Then, add or subtract.

|                | Rename.        | Add.           |                |
|----------------|----------------|----------------|----------------|

$$5\frac{3}{8}$$
$$+\,4\frac{3}{4}$$
y

$$5\frac{3}{8}$$
$$+\,4\frac{6}{8}$$
y

$$5\frac{3}{8}$$
$$+\,4\frac{6}{8}$$
$$9\frac{9}{8}$$
y

Rename.

$$9\frac{9}{8} = 10\frac{1}{8}$$

|                | Rename.        | Subtract.      |                |
|----------------|----------------|----------------|----------------|

$$8\frac{1}{6}$$
$$-\,4\frac{1}{2}$$

$$8\frac{1}{6} = 7\frac{7}{6}$$
$$4\frac{1}{2} = 4\frac{3}{6}$$

$$7\frac{7}{6}$$
$$-\,4\frac{3}{6}$$
y

$$7\frac{7}{6}$$
$$-\,4\frac{3}{6}$$
$$3\frac{4}{6}$$
y

Rename.

$$3\frac{4}{6} = 3\frac{2}{3}$$

*Add. Write the sum in simplest form.*

1. $1\frac{1}{4}$   $+\,3\frac{2}{4}$

2. $8\frac{3}{8}$   $+\,3\frac{1}{8}$

3. $13\frac{2}{5}$   $+\,7\frac{2}{5}$

4. $6\frac{3}{12}$   $+\,3\frac{5}{12}$

5. $4\frac{1}{3}$   $+\,2\frac{2}{3}$

6. $2\frac{7}{10}$   $+\,9\frac{9}{10}$

7. $24\frac{5}{7}$   $+\,15\frac{4}{7}$

8. $21\frac{1}{6}$   $+\,7\frac{5}{6}$

9. $8\frac{9}{16}$   $+\,4\frac{11}{16}$

10. $15\frac{7}{8}$   $+\,4\frac{5}{8}$

*Subtract. Write the difference in simplest form.*

11. $7\frac{7}{9}$   $-\,1\frac{3}{9}$

12. $9\frac{3}{4}$   $-\,2\frac{1}{4}$

13. $8\frac{4}{5}$   $-\,2\frac{1}{5}$

14. $17\frac{9}{10}$   $-\,6\frac{3}{10}$

15. $29\frac{5}{7}$   $-\,18\frac{2}{7}$

16. $13\frac{7}{8}$   $-\,2\frac{3}{8}$

17. $8\frac{3}{5}$   $-\,5\frac{1}{4}$

18. $7\frac{2}{3}$   $-\,2\frac{1}{5}$

19. $12\frac{11}{12}$   $-\,5\frac{1}{4}$

20. $37\frac{5}{6}$   $-\,11\frac{3}{8}$

*Solve each problem.*

21. In one year, Montreal, Quebec, had $40\frac{4}{5}$ inches rain. The same year Seattle, Washington, had $31\frac{3}{5}$ inches. What is the difference?

22. Bolivia mined $6\frac{1}{2}$ thousand tons of tin. Belgium mined $3\frac{9}{10}$ thousand tons. How much tin did the countries mine altogether.

When subtracting with mixed numerals, first subtract fractions.
Then subtract the whole numbers.

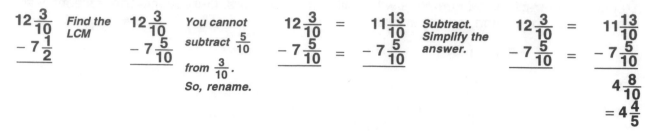

*Add or subtract. Write the sum or difference in simplest form.*

**23.** $38\frac{3}{7}$
$-\ 29\frac{9}{10}$

**24.** $6\frac{2}{3}$
$-\ 1\frac{5}{9}$

**25.** $9\frac{1}{6}$
$-\ 7\frac{1}{4}$

**26.** $55\frac{1}{3}$
$+\ 24\frac{4}{7}$

**27.** $27\frac{2}{7}$
$+\ 12\frac{3}{4}$

**28.** $56\frac{1}{3}$
$+\ 28\frac{5}{8}$

**29.** $84\frac{2}{5}$
$+\ 13\frac{5}{9}$

**30.** $46\frac{5}{6}$
$-\ 17\frac{5}{8}$

**31.** $27\frac{2}{3}$
$-\ 16\frac{1}{5}$

**32.** $44\frac{1}{7}$
$-\ 28\frac{3}{4}$

**33.** $32\frac{5}{8}$
$-\ 19\frac{3}{5}$

**34.** $21\frac{3}{5}$
$+\ 8\frac{4}{6}$

**35.** $55\frac{6}{7}$
$-\ 37\frac{4}{9}$

**36.** $5\frac{1}{6}$
$+\ 8\frac{5}{8}$

**37.** $3\frac{1}{4}$
$+\ 15\frac{5}{6}$

**38.** $3\frac{1}{3}$
$4\frac{1}{4}$
$+\ 6\frac{5}{6}$

**39.** $9\frac{5}{6}$
$+\ 7\frac{3}{10}$

**40.** $13\frac{1}{6}$
$8\frac{5}{12}$
$+\ 7\frac{3}{8}$

**41.** $6\frac{3}{4}$
$12\frac{2}{5}$
$+\ 5\frac{3}{10}$

**42.** $2\frac{3}{5}$
$1\frac{7}{15}$
$+\ 25\frac{3}{10}$

**43.** $45\frac{2}{9}$
$-\ 25\frac{3}{5}$

**44.** $17\frac{1}{6}$
$-\ 8\frac{5}{8}$

**45.** $30\frac{3}{4}$
$-\ 25\frac{9}{10}$

**46.** $23\frac{1}{5}$
$-\ 8\frac{1}{3}$

*Solve each problem.*

**47.** The United Kingdom mined $21\frac{1}{3}$ thousands tons of tin. Belgium mined $3\frac{9}{10}$ thousand tons. How much tin did the countries mine altogether?

**48.** Oregon has $21\frac{3}{5}$ people per square mile. Montana has $4\frac{7}{10}$ people per square mile. What is the difference?

12

# Multiplication of Fractions

To multiply two fractions, multiply the numerators. Then multiply the denominators.

**Multiply the numerators.**

$$\frac{3}{5} \times \frac{2}{5} = \boxed{\frac{3 \times 2}{5 \times 5}} = \frac{6}{25}$$

**Multiply the denominators.**

*Multiply. Write the product in simplest form.*

1. $\frac{1}{3} \times \frac{1}{4}$      2. $\frac{2}{5} \times \frac{1}{3}$      3. $\frac{1}{9} \times \frac{4}{7}$      4. $\frac{1}{4} \times \frac{3}{7}$

5. $\frac{3}{8} \times \frac{1}{7}$      6. $\frac{5}{8} \times \frac{3}{4}$      7. $\frac{2}{7} \times \frac{7}{12}$      8. $\frac{4}{7} \times \frac{3}{10}$

*There is a shortcut to simplest form when multiplying fractions.*

$$\frac{4}{9} \times \frac{3}{8} \quad = \quad \frac{\overset{1}{4}}{9} \times \frac{3}{\underset{2}{8}} \quad = \qquad\qquad\qquad \frac{\overset{1}{4}}{\underset{3}{9}} \times \frac{\overset{1}{3}}{\underset{2}{8}} \quad = \quad \frac{1}{6}$$

*Notice that this answer is in simplest form.*

**The GCF of 8 and 4 is 4.**
**Divide the 8 and 4 by 4.**

**What is the GCF of 3 and 9? _____**

*Multiply. Use the shortcut to simplest form.*

9. $\frac{6}{7} \times \frac{14}{15}$      10. $\frac{8}{15} \times \frac{5}{6}$      11. $\frac{5}{6} \times \frac{3}{8}$      12. $\frac{3}{8} \times \frac{4}{9}$

13. $\frac{7}{8} \times \frac{4}{7}$      14. $\frac{7}{16} \times \frac{8}{21}$      15. $\frac{5}{9} \times \frac{3}{5}$      16. $\frac{2}{3} \times \frac{1}{2} \times \frac{3}{4}$

17. $\frac{3}{16} \times \frac{4}{9}$      18. $\frac{3}{8} \times \frac{2}{3} \times \frac{1}{4}$      19. $\frac{3}{7} \times \frac{11}{15} \times \frac{7}{12}$      20. $\frac{2}{3} \times \frac{3}{5} \times \frac{6}{7}$

21. $\frac{3}{4} \times \frac{5}{8} \times \frac{4}{5}$      22. $\frac{4}{5} \times \frac{1}{3} \times \frac{15}{16}$      23. $\frac{8}{9} \times \frac{3}{5} \times \frac{15}{16}$      24. $\frac{9}{16} \times \frac{2}{3} \times \frac{5}{6}$

To multiply a fraction and a whole number, rename the whole number as a fraction.

$$3 \times \frac{3}{5} = \boxed{\frac{3}{1}} \times \frac{3}{5} \quad = \quad \frac{3 \times 3}{1 \times 5} = \frac{9}{5} \quad \text{or} \quad 1\frac{4}{5}$$

**Rename 3 as $\frac{3}{1}$.**

You can use the shortcut to simplify a problem.

$$\frac{7}{10} \times 5 = \frac{7}{10} \times \frac{\overset{1}{5}}{1} = \frac{7 \times 5}{\underset{2}{10} \times 1} = \frac{7}{2} \quad \text{or} \quad 3\frac{1}{2}$$

**What is the GCF of 5 and 10? _____**

*Multiply. Write the product in simplest form.*

25. $5 \times \frac{1}{8}$

26. $\frac{1}{3} \times 4$

27. $8 \times \frac{2}{7}$

28. $\frac{7}{9} \times 3$

29. $\frac{3}{4} \times 7$

30. $\frac{1}{5} \times 13$

31. $5 \times \frac{5}{8}$

32. $5 \times \frac{1}{9}$

33. $\frac{5}{8} \times 4$

34. $6 \times \frac{1}{3}$

35. $\frac{3}{8} \times 14$

36. $12 \times \frac{2}{9}$

37. $\frac{3}{4} \times 8$

38. $3 \times \frac{5}{9}$

39. $8 \times \frac{9}{10}$

40. $6 \times \frac{1}{4}$

41. $\frac{7}{15} \times 20$

42. $10 \times \frac{3}{14}$

43. $\frac{1}{6} \times \frac{2}{5} \times 20$

44. $\frac{5}{12} \times 18$

45. $\frac{2}{3} \times 9 \times \frac{5}{6}$

46. $\frac{7}{8} \times 16$

47. $3 \times 4 \times \frac{5}{6}$

48. $5 \times \frac{2}{15} \times \frac{3}{4}$

*Solve each problem.*

49. Mr. Lin's garden is $\frac{7}{10}$ vegetables. Of these vegetables, $\frac{5}{7}$ are beans. What part of the garden is planted in beans?

50. Mrs. Carlson's garden is $\frac{3}{4}$ flowers. Of these flowers, $\frac{4}{9}$ are tulips. What part of the garden is planted in tulips?

*Multiply. Use the shortcut to find the simplest form.*

51. $\frac{2}{5} \times \frac{1}{2}$

52. $\frac{1}{2} \times \frac{6}{13}$

53. $\frac{9}{15} \times \frac{10}{27}$

54. $\frac{6}{7} \times \frac{7}{10}$

55. $\frac{2}{3} \times \frac{3}{8}$

56. $\frac{2}{25} \times \frac{5}{8}$

57. $\frac{7}{15} \times \frac{5}{7}$

58. $\frac{2}{5} \times \frac{5}{8}$

59. $\frac{4}{5} \times \frac{15}{16}$

60. $\frac{3}{8} \times \frac{2}{7}$

61. $\frac{5}{8} \times \frac{4}{15}$

62. $\frac{9}{10} \times \frac{5}{6}$

63. $\frac{5}{6} \times \frac{24}{35}$

64. $\frac{5}{9} \times \frac{3}{20}$

65. $\frac{3}{10} \times \frac{5}{18}$

66. $\frac{4}{35} \times \frac{5}{16}$

*Multiply. Write the answer in simplest form.*

67. $\frac{3}{4} \times \frac{4}{5} \times \frac{5}{12}$

68. $\frac{1}{4} \times \frac{7}{6} \times \frac{24}{35}$

69. $\frac{2}{5} \times \frac{2}{3} \times \frac{21}{18}$

70. $\frac{1}{5} \times \frac{15}{20}$

71. $\frac{3}{5} \times \frac{2}{3} \times \frac{45}{50}$

72. $\frac{3}{10} \times \frac{5}{15} \times \frac{20}{25}$

73. $\frac{4}{18} \times \frac{3}{16}$

74. $\frac{5}{16} \times \frac{4}{15} \times \frac{12}{18}$

# Renaming Mixed Numerals

Rename $2\frac{3}{4}$ as a fraction as follows.

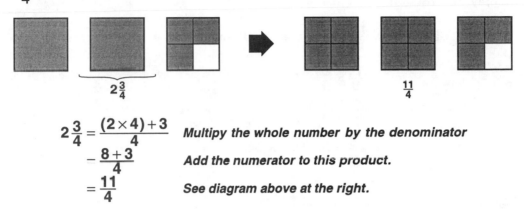

$$2\frac{3}{4} = \frac{(2 \times 4) + 3}{4}$$ **Multiply the whole number by the denominator**

$$= \frac{8 + 3}{4}$$ **Add the numerator to this product.**

$$= \frac{11}{4}$$ **See diagram above at the right.**

*Rename each mixed numeral as a fraction.*

1. $2\frac{1}{3}$      2. $3\frac{1}{2}$      3. $2\frac{3}{5}$      4. $3\frac{5}{8}$      5. $2\frac{4}{5}$

6. $3\frac{1}{6}$      7. $6\frac{3}{7}$      8. $7\frac{1}{3}$      9. $4\frac{3}{4}$      10. $5\frac{2}{3}$

11. $9\frac{9}{10}$      12. $2\frac{7}{12}$      13. $2\frac{6}{8}$      14. $3\frac{2}{3}$      15. $3\frac{5}{6}$

16. $3\frac{4}{9}$      17. $6\frac{5}{8}$      18. $9\frac{2}{3}$      19. $12\frac{1}{2}$      20. $5\frac{7}{16}$

21. $14\frac{2}{3}$      22. $10\frac{3}{10}$      23. $2\frac{13}{16}$      24. $18\frac{4}{7}$      25. $7\frac{3}{8}$

26. $8\frac{5}{9}$      27. $11\frac{3}{10}$      28. $3\frac{5}{12}$      29. $7\frac{4}{15}$      30. $11\frac{4}{5}$

31. $14\frac{2}{5}$      32. $6\frac{1}{15}$      33. $2\frac{11}{12}$      34. $2\frac{3}{4}$      35. $2\frac{1}{9}$

36. $1\frac{3}{5}$      37. $1\frac{6}{7}$      38. $5\frac{7}{12}$      39. $9\frac{3}{8}$      40. $16\frac{1}{3}$

41. $2\frac{8}{9}$      42. $4\frac{1}{10}$      43. $20\frac{5}{6}$      44. $6\frac{6}{7}$      45. $2\frac{1}{3}$

*Solve each problem.*

46. In December, Minneapolis had $3\frac{7}{20}$ inches of snow. In January, Minneapolis had $7\frac{3}{25}$ inches of snow. How many inches of snow did Minneapolis have in December and January?

47. In May, Savannah, Georgia, had $3\frac{1}{5}$ inches rain. In April, Savannah had $2\frac{2}{5}$ inches. What is the difference?

# Multiplication with Mixed Numerals

To multiply with mixed numerals, first rename the mixed numerals as fractions.
Then, multiply the fractions.

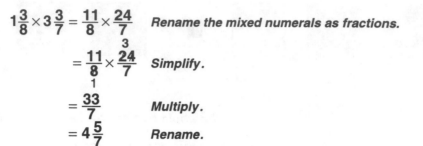

$$1\frac{3}{8} \times 3\frac{3}{7} = \frac{11}{8} \times \frac{24}{7}$$ *Rename the mixed numerals as fractions.*

$$= \frac{11}{8} \times \frac{\overset{3}{24}}{7}$$ *Simplify.*

$$= \frac{33}{7}$$ *Multiply.*

$$= 4\frac{5}{7}$$ *Rename.*

*Complete.*

**1.** $9 \times \frac{5}{6} = \frac{9}{1} \times \frac{5}{6}$

**2.** $5\frac{1}{3} \times 1\frac{4}{5} = \frac{16}{3} \times \frac{9}{5}$

**3.** $\frac{1}{2} \times 4\frac{2}{5} = \frac{1}{2} \times \frac{22}{5}$

*Multiply. Write the product in simplest form.*

**4.** $\frac{1}{3} \times 1\frac{1}{7}$

**5.** $\frac{2}{7} \times 4\frac{1}{6}$

**6.** $\frac{4}{5} \times 5\frac{1}{4}$

**7.** $2\frac{3}{5} \times \frac{5}{6}$

**8.** $1\frac{5}{6} \times \frac{5}{9}$

**9.** $3\frac{1}{4} \times \frac{2}{5}$

**10.** $4\frac{3}{4} \times \frac{1}{3}$

**11.** $7\frac{5}{8} \times \frac{4}{5}$

**12.** $\frac{5}{8} \times 5$

**13.** $\frac{5}{6} \times 8$

**14.** $\frac{3}{5} \times 20$

**15.** $7 \times \frac{2}{3}$

**16.** $1\frac{3}{4} \times 7$

**17.** $1\frac{1}{5} \times 4$

**18.** $8\frac{2}{3} \times 6$

**19.** $3\frac{1}{5} \times 6$

**20.** $\frac{7}{8} \times 1\frac{3}{7}$

**21.** $\frac{2}{5} \times 1\frac{1}{4}$

**22.** $1\frac{1}{3} \times \frac{9}{10}$

**23.** $1\frac{3}{5} \times \frac{3}{4}$

**24.** $4\frac{3}{8} \times 2$

**25.** $7 \times 3\frac{1}{2}$

**26.** $4 \times 1\frac{1}{3}$

**27.** $2\frac{1}{8} \times 1\frac{1}{3}$

**28.** $1\frac{1}{4} \times 1\frac{3}{5}$

**29.** $1\frac{3}{4} \times 2$

**30.** $1\frac{5}{7} \times 2\frac{5}{8}$

**31.** $1\frac{1}{5} \times 3\frac{3}{4}$

*Solve.*

**32.** A dracaena is $4\frac{7}{8}$ times as tall as a wax begonia. A wax begonia is $1\frac{1}{6}$ feet tall. How tall is a dracaena?

**33.** In a $\frac{9}{16}$-ounce bottle of nail polish, there is $\frac{3}{8}$-ounce water. What part of the nail polish is water?

# Division of Fractions

Two numbers whose product is 1 are called **reciprocals**.

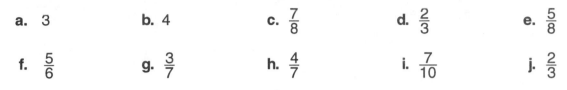

*Name the reciprocals for each of the following.*

**a.** 3 **b.** 4 **c.** $\frac{7}{8}$ **d.** $\frac{2}{3}$ **e.** $\frac{5}{8}$

**f.** $\frac{5}{6}$ **g.** $\frac{3}{7}$ **h.** $\frac{4}{7}$ **i.** $\frac{7}{10}$ **j.** $\frac{2}{3}$

Reciprocals are used in division of fractions as shown below.

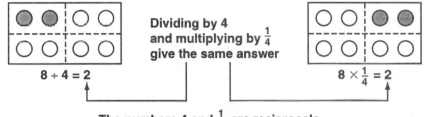

The numbers 4 and $\frac{1}{4}$ are reciprocals.

Therefore, to divide by a fraction, multiply by its reciprocal.

$$\frac{3}{8} \div \frac{1}{4} = \frac{3}{8} \times \frac{4}{1} = \frac{3}{8} \times \frac{\overset{1}{4}}{1} = \frac{3}{2} \text{ or } 1\frac{1}{2}$$

*Copy and complete.*

**1.** $3 \div \frac{2}{5} = \frac{3}{1} \times \frac{5}{2} = $ _____

**2.** $4 \div \frac{1}{3} = \frac{4}{1} \times \frac{3}{1} = $ _____

**3.** $\frac{3}{8} \div \frac{3}{4} = \frac{3}{8} \times \frac{4}{3} = $ _____

**4.** $\frac{1}{8} \div ? = \frac{1}{8} \times \frac{1}{2} = $ _____

**5.** $\frac{2}{5} \div \frac{7}{10} = \frac{2}{5} \times \frac{\square}{\square} = $ _____

**6.** $\frac{1}{4} \div \frac{3}{5} = \frac{1}{4} \times \frac{\square}{\square} = $ _____

*Divide. Write the quotient in simplest form.*

**7.** $6 \div \frac{1}{6}$    **8.** $5 \div \frac{1}{3}$    **9.** $4 \div \frac{1}{3}$    **10.** $7 \div \frac{1}{4}$

**11.** $8 \div \frac{2}{5}$    **12.** $10 \div \frac{2}{7}$    **13.** $12 \div \frac{6}{11}$    **14.** $12 \div \frac{2}{3}$

**15.** $\frac{1}{2} \div 5$    **16.** $\frac{1}{4} \div 12$    **17.** $\frac{1}{3} \div 6$    **18.** $\frac{1}{7} \div 2$

**19.** $\frac{9}{15} \div \frac{3}{15}$    **20.** $\frac{7}{8} \div \frac{7}{8}$    **21.** $\frac{4}{7} \div \frac{9}{28}$    **22.** $\frac{4}{9} \div \frac{4}{5}$

**23.** $\frac{1}{4} \div \frac{5}{8}$    **24.** $\frac{1}{7} \div \frac{5}{6}$    **25.** $\frac{1}{2} \div \frac{5}{12}$    **26.** $\frac{1}{16} \div \frac{3}{4}$

27. $\frac{3}{4} \div \frac{3}{8}$

28. $\frac{3}{5} \div \frac{3}{4}$

29. $\frac{7}{8} \div \frac{3}{8}$

30. $8 \div \frac{1}{3}$

31. $6 \div \frac{4}{5}$

32. $9 \div \frac{5}{8}$

33. $5 \div \frac{2}{5}$

34. $\frac{5}{12} \div \frac{2}{3}$

35. $7 \div \frac{1}{2}$

36. $6 \div \frac{3}{4}$

37. $\frac{9}{10} \div 3$

38. $\frac{2}{3} \div \frac{3}{4}$

39. $5 \div \frac{1}{8}$

40. $2 \div \frac{5}{6}$

41. $\frac{7}{10} \div \frac{14}{25}$

42. $\frac{4}{7} \div \frac{1}{3}$

43. $\frac{2}{7} \div 5$

44. $\frac{5}{6} \div 2$

45. $\frac{5}{8} \div \frac{5}{6}$

46. $\frac{4}{5} \div 4$

47. $\frac{1}{6} \div \frac{2}{3}$

48. $\frac{7}{10} \div \frac{2}{3}$

49. $\frac{7}{9} \div \frac{7}{12}$

50. $\frac{3}{4} \div \frac{6}{7}$

51. $\frac{8}{9} \div \frac{4}{9}$

52. $\frac{5}{8} \div \frac{5}{6}$

53. $\frac{7}{20} \div \frac{28}{35}$

54. $\frac{11}{16} \div \frac{3}{8}$

*Solve each problem.*

55. A stingless bee is $\frac{1}{12}$ inch long. This is $\frac{2}{9}$ of the length of a dwarf bee. How long is the dwarf bee?

56. A worker honeybee collects enough nectar in its lifetime to make about $\frac{1}{10}$ pound of honey. How many bees are needed to make $\frac{4}{5}$ pound of honey?

57. In $\frac{3}{5}$ ounce of fertilizer, there is $\frac{3}{10}$ ounce nitrogen. What part of the fertilizer is nitrogen?

58. In 4 ounces of fertilizer, there is $\frac{2}{3}$ ounce phosphorus. What part of the fertilizer is phosphorus?

59. A honeybee collects enough nectar to make about $\frac{1}{10}$ pound of honey. How many bees are needed to make $\frac{3}{5}$ pound of honey?

60. To make a model airplane, Lane cut a $\frac{5}{6}$ inch piece of balsa wood into $\frac{1}{12}$ inch strips. How many strips did she have?

# Division with Mixed Numerals

To divide with mixed numerals, rename them as fractions.

$$10\tfrac{1}{2} \div 2\tfrac{1}{2} = \frac{21}{2} \div \frac{5}{2}$$ **Rename the mixed numerals as fractions.**

$$= \frac{21}{2} \times \frac{2}{5}$$ **Multiply by the reciprocal of $\frac{5}{2}$.**

$$= \frac{21}{2} \times \frac{\overset{1}{2}}{5}$$ **Simplify.**

$$= \frac{21}{5}$$

$$= 4\tfrac{1}{5}$$

*Copy and complete.*

1. $1\tfrac{3}{5} \div 2\tfrac{2}{5} = \frac{8}{5} \div \frac{12}{5} = \frac{8}{5} \times \frac{5}{12} =$ _____

2. $2\tfrac{5}{8} \div 7\tfrac{1}{2} = \frac{21}{8} \div \frac{15}{2} = \frac{21}{8} \times \frac{2}{15} =$ _____

3. $4 \div 2\tfrac{1}{3} = \frac{4}{1} \div \frac{7}{3} = \frac{4}{1} \times \frac{\square}{\square} =$ _____

4. $3\tfrac{3}{5} \div 6 = \frac{18}{5} \div \frac{6}{1} = \frac{18}{5} \times \frac{\square}{\square} =$ _____

5. $1\tfrac{1}{5} \div 1\tfrac{3}{5} = \frac{6}{5} \div \frac{8}{5} = \frac{6}{5} \times \frac{\square}{\square} =$ _____

6. $2\tfrac{1}{4} \div 1\tfrac{7}{8} = \frac{9}{4} \div \frac{15}{8} = \frac{9}{4} \times \frac{\square}{\square} =$ _____

*Divide. Write each quotient in simplest form.*

7. $5 \div 3\tfrac{1}{3}$

8. $6\tfrac{1}{4} \div 5$

9. $4 \div 1\tfrac{1}{3}$

10. $1\tfrac{5}{8} \div 2$

11. $1\tfrac{3}{5} \div \tfrac{5}{8}$

12. $3\tfrac{1}{8} \div 1\tfrac{1}{4}$

13. $3 \div 4\tfrac{1}{2}$

14. $2\tfrac{3}{4} \div \tfrac{5}{6}$

15. $\tfrac{5}{6} \div 1\tfrac{1}{9}$

16. $2\tfrac{2}{5} \div \tfrac{3}{10}$

17. $\tfrac{8}{9} \div 2\tfrac{2}{5}$

18. $1\tfrac{2}{5} \div 2\tfrac{2}{3}$

19. $\tfrac{9}{10} \div 5\tfrac{2}{5}$

20. $1\tfrac{1}{4} \div 4\tfrac{1}{2}$

21. $4\tfrac{2}{3} \div 1\tfrac{3}{5}$

22. $3\tfrac{1}{2} \div 5\tfrac{1}{2}$

23. $3\tfrac{1}{5} \div 1\tfrac{1}{3}$

24. $1\tfrac{1}{4} \div 7\tfrac{1}{2}$

25. $4\tfrac{3}{4} \div 1\tfrac{7}{8}$

26. $5\tfrac{1}{2} \div 2\tfrac{3}{4}$

27. $9\tfrac{3}{5} \div 4\tfrac{1}{5}$

**28.** $1\frac{2}{3} \div 6\frac{3}{7}$      **29.** $2\frac{3}{5} \div 11\frac{4}{5}$      **30.** $2\frac{1}{9} \div 3\frac{1}{3}$

**31.** $2\frac{1}{4} \div 2\frac{1}{4}$      **32.** $1\frac{1}{8} \div 3\frac{3}{8}$      **33.** $4\frac{5}{9} \div 1\frac{2}{9}$

**34.** $1\frac{1}{9} \div 2\frac{4}{9}$      **35.** $1\frac{1}{5} \div 2\frac{3}{5}$      **36.** $6\frac{3}{5} \div 2\frac{1}{5}$

**37.** $3\frac{9}{10} \div 2\frac{13}{15}$      **38.** $10\frac{1}{7} \div 3\frac{4}{7}$      **39.** $2\frac{2}{7} \div 2\frac{1}{2}$

**40.** $5\frac{1}{21} \div 2\frac{21}{35}$      **41.** $3\frac{2}{7} \div 2\frac{9}{21}$      **42.** $3\frac{3}{5} \div 2\frac{10}{15}$

**43.** $5\frac{1}{3} \div 1\frac{1}{9}$      **44.** $1\frac{3}{8} \div 3\frac{1}{16}$      **45.** $3\frac{5}{8} \div 7\frac{11}{16}$

**46.** $7\frac{1}{4} \div 3\frac{7}{8}$      **47.** $2\frac{2}{9} \div 1\frac{7}{18}$      **48.** $2\frac{2}{3} \div 1\frac{1}{15}$

**49.** $10 \div 2\frac{1}{5}$      **50.** $2\frac{2}{3} \div 4$      **51.** $3 \div 1\frac{1}{8}$

**52.** $\frac{7}{8} \div 1\frac{1}{4}$      **53.** $1\frac{1}{6} \div \frac{2}{3}$      **54.** $\frac{9}{10} \div 2\frac{4}{5}$

**55.** $3\frac{1}{5} \div 1\frac{1}{3}$      **56.** $1\frac{2}{9} \div 3\frac{1}{5}$      **57.** $6\frac{1}{2} \div 4\frac{1}{4}$

*Solve each problem.*

**58.** A $3\frac{1}{3}$-foot row yields 1 pound of spinach. How many pounds per foot does each row yield?

**59.** A $6\frac{2}{3}$-foot row yields $5\frac{1}{3}$ pounds of swiss chard. How many pounds per foot does each row yield?

**60.** There are $3\frac{1}{2}$ pounds of oranges to be divided among 5 children. How much does each receive?

**61.** A dolphin swims $2\frac{7}{9}$ miles in $6\frac{2}{3}$ minutes. How fast can the dolphin swim per minute?

**62.** An owl flies $2\frac{2}{9}$ miles in $3\frac{1}{3}$ minutes. How fast can the owl fly per minute?

**63.** A $5\frac{3}{4}$-foot row yields $6\frac{1}{2}$ pounds of peas. How many pounds per foot does each row yield?

**64.** A dragonfly flies 3 miles in $3\frac{3}{5}$ minutes. How fast can the dragonfly fly per minute?

**65.** A whale swims $1\frac{5}{6}$ miles in $5\frac{1}{2}$ minutes. How fast can the whale swim per minute?

# Circumference of a Circle

The diameter (*d*) of a circle is
twice the length of the radius (*r*).

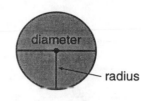
diameter
radius

The distance around a circle is called the **circumference**.

For all circles, the ratio of the circumference to the diameter is the same number.
The Greek letter $\pi$(pi) stands for this number.

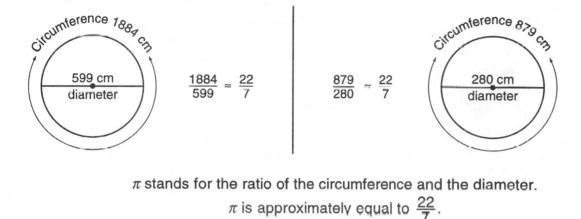

Circumference 1884 cm

599 cm
diameter

$$\frac{1884}{599} \approx \frac{22}{7}$$

$$\frac{879}{280} \approx \frac{22}{7}$$

Circumference 879 cm

280 cm
diameter

$\pi$ stands for the ratio of the circumference and the diameter.

$\pi$ is approximately equal to $\frac{22}{7}$.

To find the circumference, multiply $\pi$(pi) by
the length of the diameter ($C = \pi \times d$).

14 ft
diameter

$$C \approx \frac{22}{7} \times 14 \approx 44$$

**The circumference is 44 feet.**

Another formula for circumference is $C = 2 \times \pi \times r$.

*Find the circumference of each circle described below. Use $\frac{22}{7}$ for $\pi$.*

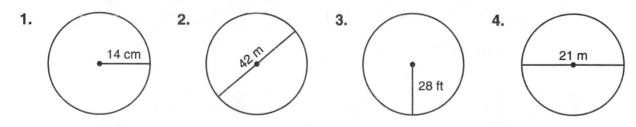

1.   14 cm

2.   42 m

3.   28 ft

4.   21 m

*Find the circumference of each circle whose diameter is listed below. Use $\frac{22}{7}$ for $\pi$.*

| | | |
|---|---|---|
| **5.** 63 ft | **6.** 77 m | **7.** 126 mm |
| **8.** 1484 ft | **9.** 231 cm | **10.** 448 in. |
| **11.** 287 cm | **12.** 182 in. | **13.** 47 in. |
| **14.** 9 ft | **15.** 5 in. | **16.** 12 cm |

# Addition of Integers

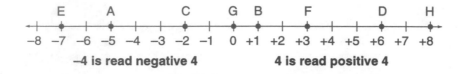

-4 is read negative 4          4 is read positive 4

The set of numbers shown on the number line above is called **integers**. The set of integers includes positive and negative numbers, and zero, which is neither negative nor positive.

*Name the integer for each letter on the number line above.*

**1.** E          **2.** B          **3.** D          **4.** F

**5.** C          **6.** H          **7.** A          **8.** G

The diagrams on the number lines below show addition of integers.

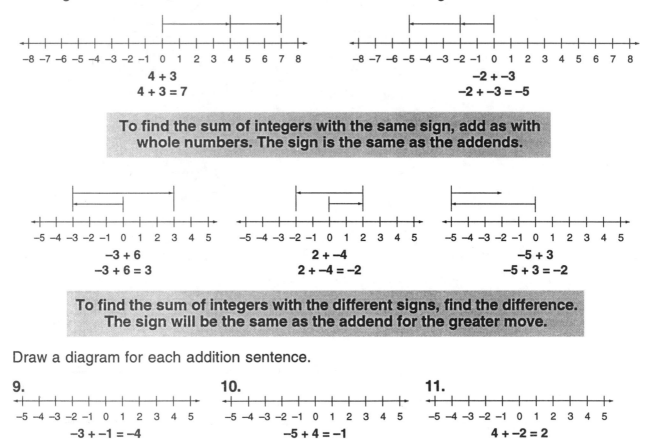

To find the sum of integers with the same sign, add as with whole numbers. The sign is the same as the addends.

To find the sum of integers with the different signs, find the difference. The sign will be the same as the addend for the greater move.

Draw a diagram for each addition sentence.

**9.**
-5 -4 -3 -2 -1 0 1 2 3 4 5
-3 + -1 = -4

**10.**
-5 -4 -3 -2 -1 0 1 2 3 4 5
-5 + 4 = -1

**11.**
-5 -4 -3 -2 -1 0 1 2 3 4 5
4 + -2 = 2

*Add.*

**12.** -7 + -3 =          **13.** -7 + 5 =          **14.** 9 + 12 =          **15.** 8 + -8 =

**16.** 0 + -9 =          **17.** -5 + 9 =          **18.** -4 + -11 =          **19.** 9 + -7 =

**20.** -10 + 8 =          **21.** 4 + -4 =          **22.** -7 + 0 =          **23.** -4 + -6 =

# Subtraction of Integers

The diagram on the number line shows opposite integers.

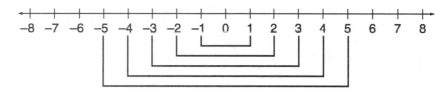

Use opposites to solve subtraction problems.

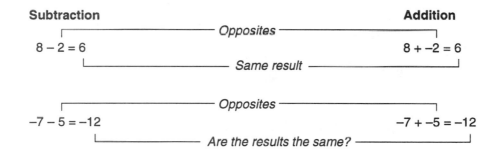

> **To subtract an integer, add its opposite.**

*Write the opposite of each integer.*

1. 7
2. −7
3. −2
4. 8
5. −11

6. −5
7. 14
8. −12
9. −17
10. 1

*Write an addition problem for each of the following.*

11. $7 - 5$
12. $-9 - 6$
13. $6 - -4$

14. $-8 - 3$
15. $3 - -2$
16. $0 - -9$

*Subtract.*

17. $0 - -2 =$
18. $-7 - -9 =$
19. $9 - 11 =$
20. $-8 - -8 =$

21. $6 - 5 =$
22. $-8 - 6 =$
23. $5 - -4 =$
24. $-7 - 3 =$

25. $4 - -1 =$
26. $-11 - 6 =$
27. $-3 - -11 =$
28. $14 - 8 =$

29. $4.8 - 2.9 =$
30. $7 - -8 =$
31. $-5.4 - 3.7 =$
32. $-3 - 5 =$

# Multiplication of Integers

Imagine a tank being filled at the rate of 4 gallons per minute.

In 6 minutes, the tank will contain 24 more gallons.

$6 \times 4 = 24$

6 minutes ago, the tank had 24 fewer gallons.

$-6 \times 4 = -24$

Imagine a tank being emptied at the rate of 4 gallons per minute.

In 6 minutes, the tank will contain 24 less gallons.

$6 \times -4 = -24$

6 minutes ago, the tank had 24 more gallons.

$-6 \times -4 = 24$

What is the sign of the product if both integers are positive?

What is the sign of the product if both integers are negative?

**The product of two integers with the same sign is positive.**

What is the sign of the product of two integers with different signs?

**The product of two integers with different signs is negative.**

*Multiply.*

**1.** $4 \times 4 =$    **2.** $-4 \times -2 =$    **3.** $9 \times 3 =$    **4.** $-8 \times -6 =$

**5.** $-4 \times 2 =$    **6.** $11 \times -3 =$    **7.** $7 \times -5 =$    **8.** $-7 \times 9 =$

**9.** $-4 \times -7 =$    **10.** $-5 \times 5 =$    **11.** $7 \times -7 =$    **12.** $3 \times 5 =$

**13.** $2 \times -6 =$    **14.** $-9 \times -4 =$    **15.** $3 \times 2 =$    **16.** $-3 \times 8 =$

**17.** $-14 \times 10 =$    **18.** $16 \times 21 =$    **19.** $8 \times -17 =$    **20.** $-11 \times -15 =$

**21.** $25 \times 25 =$    **22.** $-32 \times 30 =$    **23.** $-30 \times -10 =$    **24.** $70 \times -25 =$

# Division of Integers

Division undoes multiplication.
You can learn about the division of integers by studying this relationship.

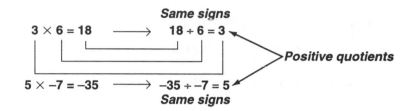

What is the sign of the quotient if the signs of both integers are the same?

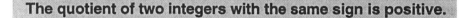

**The quotient of two integers with the same sign is positive.**

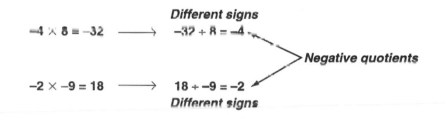

What is the sign of the quotient if the sign of both integers are different?

**The quotient of two integers with different signs is negative.**

*Divide.*

**1.** $32 \div 2 =$      **2.** $-40 \div -2 =$      **3.** $18 \div 6 =$      **4.** $-21 \div -3 =$

**5.** $42 \div -6 =$      **6.** $-24 \div 4 =$      **7.** $81 \div -3 =$      **8.** $-40 \div 10 =$

**9.** $-18 \div -6 =$      **10.** $-81 \div 9 =$      **11.** $36 \div 4 =$      **12.** $90 \div -9 =$

**13.** $-25 \div 5 =$      **14.** $64 \div 8 =$      **15.** $64 \div -2 =$      **16.** $-140 \div -5 =$

**17.** $-96 \div -12 =$      **18.** $-48 \div 8 =$      **19.** $72 \div 6 =$      **20.** $72 \div -8 =$

**21.** $1296 \div -54 =$      **22.** $-852 \div -71 =$      **23.** $-660 \div 44 =$      **24.** $1512 \div 27 =$

# ANSWERS and SOLUTIONS for RATIONALS

## Decimals to Fractions (page 1)

1. 0.6, $\frac{6}{10}$  2. 0.1, $\frac{1}{10}$  3. 0.3, $\frac{3}{10}$  4. 0.2, $\frac{2}{10}$  5. 0.25, $\frac{25}{100}$  6. 0.3, $\frac{3}{10}$  7. 0.20, $\frac{20}{100}$  8. 0.34, $\frac{34}{100}$

9. 0.308, $\frac{308}{1000}$  10. 0.7, $\frac{7}{10}$  11. 3.3, $3\frac{3}{10}$  12. 0.60, $\frac{60}{100}$  13. 0.09, $\frac{9}{100}$  14. 2.06, $2\frac{6}{100}$

15. 0.002, $\frac{2}{1000}$  16. 0.02, $\frac{2}{100}$  17. 1.0, $\frac{10}{10}$ or 1  18. 1.6, $1\frac{6}{10}$  19. 0.605, $\frac{605}{1000}$  20. 0.81, $\frac{81}{100}$

21. 0.019, $\frac{19}{1000}$  22. 3.857, $3\frac{857}{1000}$  23. 7.3, $7\frac{3}{10}$  24. 0.023, $\frac{23}{1000}$  25. 0.02, $\frac{2}{100}$

26. thirty-two and forty-seven hundredths  27. $\frac{9}{10}$  28. eighty-three hundredths $= \frac{83}{100}$

29. sixth and seven tenths $= 6\frac{7}{10}$  30. one hundred seventy-one thousandths $= \frac{171}{1000}$

31. one and seventy-five thousandths $= 1\frac{75}{1000}$  32. five and six hundredths $= 5\frac{6}{100}$

33. five hundred forty-one thousandths $= \frac{541}{1000}$  34. twelve and nine hundredths $= 12\frac{9}{100}$

35. one thousandth $= \frac{1}{1000}$

## Greatest Common Factor (page 2)

1. 9 y 1, 3, 9  2. 8 y 1, 2, 4, 8  3. 12 y 1, 2, 3, 4, 6, 12  4. 15 y 1, 3, 5, 15  5. 30 y 1, 2, 3, 5, 6, 10, 15, 30

6. 18 y 1, 2, 3, 6, 9, 18

7. $8 = \boxed{2} \times 2 \times 2$  GCF = 2
$10 = \boxed{2} \times 5$

8. $\boxed{2} = 2$  GCF = 2
$10 = \boxed{2} \times 5$

9. $9 = \boxed{3} \times 3$  GCF = 3
$12 = \boxed{3} \times 2 \times 2$

10. $5 = \boxed{5}$  GCF = 5
$10 = 2 \times \boxed{5}$

11. $3 = \boxed{1} \times 3$  GCF = 1
$7 = \boxed{1} \times 7$

12. $6 = \boxed{2 \times 3}$  GCF = 6
$18 = \boxed{2 \times 3} \times 3$

13. $5 = \boxed{5}$  GCF = 5
$25 = \boxed{5} \times 5$

14. $4 = \boxed{2 \times 2}$  GCF = 4
$12 = \boxed{2 \times 2} \times 3$

15. $\frac{9}{12} = \frac{3 \times \boxed{3}}{2 \times 2 \times \boxed{3}}$  GCF = 3

16. $\frac{6}{9} = \frac{2 \times \boxed{3}}{3 \times \boxed{3}}$  GCF = 3

17. $\frac{9}{15} = \frac{\boxed{3} \times 3}{\boxed{3} \times 5}$  GCF = 3

18. $\frac{14}{16} = \frac{\boxed{2} \times 7}{\boxed{2} \times 2 \times 2 \times 2}$  GCF = 2

19. $\frac{12}{18} = \frac{2 \times \boxed{2 \times 3}}{\boxed{2 \times 3} \times 3}$  GCF = 6

20. $\frac{10}{12} = \frac{\boxed{2} \times 5}{\boxed{2} \times 2 \times 3}$  GCF = 2

21. $\frac{15}{20} = \frac{3 \times \boxed{5}}{2 \times 2 \times \boxed{5}}$  GCF = 5

22. $\frac{3}{18} = \frac{\boxed{3}}{2 \times 3 \times \boxed{3}}$  GCF = 3

23. $\frac{10}{30} = \frac{\boxed{2 \times 5}}{\boxed{2 \times 5} \times 3}$  GCF = 10

24. $\frac{4}{38} = \frac{\boxed{2} \times 2}{\boxed{2} \times 19}$  GCF = 2

25. $\frac{25}{55} = \frac{\boxed{5} \times 5}{\boxed{5} \times 11}$  GCF = 5

26. $\frac{9}{48} = \frac{3 \times \boxed{3}}{2 \times 2 \times 2 \times 2 \times \boxed{3}}$  GCF = 3

27. $\frac{16}{24} = \frac{\boxed{2 \times 2 \times 2} \times 2}{\boxed{2 \times 2 \times 2} \times 3}$  GCF = 8

28. $\frac{40}{50} = \frac{2 \times 2 \times \boxed{2 \times 5}}{\boxed{2 \times 5} \times 5}$  GCF = 10

29. $\frac{63}{72} = \frac{\boxed{3 \times 3} \times 7}{2 \times 2 \times 2 \times \boxed{3 \times 3}}$  GCF = 9

30. $\frac{42}{66} = \frac{\boxed{2 \times 3} \times 7}{\boxed{2 \times 3} \times 11}$  GCF = 6

31. $\frac{39}{51} = \frac{\boxed{3} \times 13}{\boxed{3} \times 17}$  GCF = 3

32. $\frac{45}{105} = \frac{3 \times \boxed{3 \times 5}}{\boxed{3 \times 5} \times 7}$  GCF = 15

33. $\frac{26}{72} = \frac{\boxed{2} \times 13}{\boxed{2} \times 2 \times 2 \times 3 \times 3}$  GCF = 2

34. $\frac{84}{90} = \frac{2 \times \boxed{2 \times 3} \times 7}{\boxed{2 \times 3} \times 3 \times 5}$  GCF = 6

## Simplifying Fractions (page 3)

1. $\frac{1}{2}$  2. $\frac{1}{3}$  3. $\frac{1}{3}$  4. $\frac{1}{4}$  5. $\frac{3}{5}$  6. $\frac{1}{3}$  7. $\frac{2}{3}$  8. $\frac{3}{8}$  9. $\frac{3}{4}$  10. $\frac{7}{9}$  11. $\frac{2}{3}$  12. $\frac{3}{4}$  13. $\frac{3}{5}$

14. $\frac{3}{4}$  15. $\frac{2}{3}$  16. $\frac{2}{5}$  17. $\frac{6}{7}$  18. $\frac{5}{6}$  19. $\frac{5}{8}$  20. $\frac{3}{8}$  21. $\frac{5}{6}$  22. $\frac{2}{9}$  23. $\frac{2}{5}$  24. $\frac{3}{7}$  25. $\frac{3}{10}$

26. $\frac{7}{8}$  27. $\frac{1}{7}$  28. $\frac{2}{3}$  29. $\frac{5}{6}$  30. $\frac{2}{5}$  31. $\frac{3}{5}$  32. $\frac{2}{9}$  33. $\frac{7}{11}$  34. $\frac{1}{5}$  35. $\frac{2}{3}$  36. $\frac{4}{7}$  37. $\frac{1}{4}$

38. $\frac{8}{9}$  39. $\frac{3}{4}$  40. $\frac{9}{16}$

## Improper Fractions (page 4)

1. $9 \div 4$  2. $8 \div 3$  3. $17 \div 5$  4. $16 \div 10$  5. $19 \div 9$  6. $7 \div 4$  7. $36 \div 3$  8. $169 \div 13$  9. $60 \div 3$  10. $12 \div 4$

11. $59 \div 13$  12. $131 \div 19$  13. $1\frac{5}{4}$  14. $1\frac{1}{8}$  15. $1\frac{4}{9}$  16. $1\frac{1}{3}$  17. $2\frac{1}{3}$  18. $8$  19. $2\frac{1}{7}$  20. $5$  21. $1\frac{1}{6}$

22. $3\frac{1}{3}$  23. $1\frac{2}{17}$  24. $3\frac{1}{7}$  25. $1\frac{3}{5}$  26. $1\frac{2}{3}$  27. $3$  28. $1\frac{1}{3}$  29. $1\frac{1}{2}$  30. $2\frac{1}{2}$  31. $2\frac{3}{4}$  32. $1\frac{1}{2}$

33. $9$  34. $1\frac{1}{4}$  35. $8$  36. $3\frac{1}{2}$  37. $1\frac{2}{3}$  38. $1\frac{1}{5}$  39. $1\frac{1}{3}$  40. $1\frac{2}{5}$  41. $3\frac{2}{3}$  42. $4\frac{1}{2}$  43. $2\frac{1}{2}$

44. $1\frac{11}{12}$  45. $7\frac{1}{4}$  46. $2\frac{2}{3}$  47. $7\frac{5}{6}$  48. $5\frac{1}{3}$  49. $2\frac{2}{5}$  50. $1\frac{1}{2}$  51. $1\frac{1}{4}$  52. $1\frac{1}{2}$  53. $1\frac{2}{7}$  54. $4\frac{1}{2}$

55. $2\frac{3}{4}$  56. $1\frac{2}{5}$  57. $1\frac{1}{2}$  58. $2\frac{3}{4}$  59. $3\frac{2}{3}$  60. $2\frac{1}{2}$

## Addition and Subtraction of Fractions (page 5)

1. $\frac{1}{5}+\frac{4}{5}=\frac{1+4}{5}=\frac{5}{5}$ or $1$

2. $\frac{3}{8}+\frac{7}{8}=\frac{3+7}{8}=\frac{10}{8}$ or $1\frac{1}{4}$

3. $\frac{9}{10}+\frac{1}{10}=\frac{9+1}{10}=\frac{10}{10}$ or $1$

4. $\frac{4}{7}+\frac{3}{7}=\frac{4+3}{7}=\frac{7}{7}$ or $1$

5. $\frac{7}{9}-\frac{4}{9}=\frac{7-4}{9}=\frac{3}{9}$ or $\frac{1}{3}$

6. $\frac{13}{15}-\frac{9}{15}=\frac{13-9}{15}=\frac{4}{15}$

7. $\frac{3}{7}-\frac{1}{7}=\frac{3-1}{7}=\frac{2}{7}$

8. $\frac{10}{11}-\frac{9}{11}=\frac{10-9}{11}=\frac{1}{11}$

9. $\frac{3}{4}+\frac{1}{4}=\frac{3+1}{4}=\frac{4}{4}$ or $1$

10. $\frac{4}{6}+\frac{2}{6}=\frac{4+2}{6}=\frac{6}{6}$ or $1$

11. $\frac{7}{8}+\frac{2}{8}=\frac{7+2}{8}=\frac{9}{8}$ or $1\frac{1}{8}$

12. $\frac{4}{9}+\frac{4}{9}=\frac{4+4}{9}=\frac{8}{9}$

13. $\frac{6}{18}-\frac{3}{18}=\frac{6-3}{18}=\frac{3}{18}$ or $\frac{1}{6}$

14. $\frac{8}{16}-\frac{3}{16}=\frac{8-3}{16}=\frac{5}{16}$

15. $\frac{15}{30}-\frac{8}{30}=\frac{15-8}{30}=\frac{7}{30}$

16. $\frac{15}{35}-\frac{13}{35}=\frac{15-13}{35}=\frac{2}{35}$

17. $\frac{9}{20}+\frac{3}{20}=\frac{9+3}{20}=\frac{12}{20}$ or $\frac{3}{5}$

18. $\frac{4}{18}+\frac{14}{18}=\frac{4+14}{18}=\frac{18}{18}$ or $1$

19. $\frac{4}{10}+\frac{5}{10}=\frac{4+5}{10}=\frac{9}{10}$

20. $\frac{3}{8}+\frac{5}{8}=\frac{3+5}{8}=\frac{8}{8}$ or $1$

21. $\frac{17}{18}-\frac{5}{18}=\frac{17-5}{18}=\frac{12}{18}$ or $\frac{2}{3}$

22. $\frac{4}{7}+\frac{6}{7}=\frac{4+6}{7}=\frac{10}{7}$ or $1\frac{3}{7}$

23. $\frac{20}{21}-\frac{8}{21}=\frac{20-8}{21}=\frac{12}{21}$ or $\frac{4}{7}$

24. $\frac{5}{9}+\frac{8}{9}=\frac{5+8}{9}=\frac{13}{9}$ or $1\frac{4}{9}$

25. $\frac{29}{36}-\frac{11}{36}=\frac{29-11}{36}=\frac{18}{36}$ or $\frac{1}{2}$

26. $\frac{2}{3}+\frac{2}{3}=\frac{2+2}{3}=\frac{4}{3}$ or $1\frac{1}{3}$

27. $\frac{15}{32}-\frac{10}{32}=\frac{15-10}{32}=\frac{5}{32}$

28. $\frac{4}{15}-\frac{1}{15}=\frac{4-1}{15}=\frac{3}{15}$ or $\frac{1}{5}$

29. $\frac{5}{11}-\frac{1}{11}=\frac{5-1}{11}=\frac{4}{11}$

30. $\frac{3}{4}+\frac{3}{4}=\frac{3+3}{4}=\frac{6}{4}$ or $1\frac{1}{2}$

31. $\frac{8}{15}+\frac{11}{15}=\frac{8+11}{15}=\frac{19}{15}$ or $1\frac{4}{15}$

32. $\frac{7}{10}-\frac{1}{10}=\frac{7-1}{10}=\frac{6}{10}$ or $\frac{3}{5}$

33. $\frac{23}{38}-\frac{9}{28}=\frac{23-9}{28}=\frac{14}{28}$ or $\frac{1}{2}$

34. $\frac{11}{24}+\frac{23}{24}=\frac{11+23}{24}=\frac{34}{24}$ or $1\frac{5}{12}$

35. $\frac{8}{9}-\frac{2}{9}=\frac{8-2}{9}=\frac{6}{9}$ or $\frac{2}{3}$

36. $\frac{10}{21}+\frac{17}{21}=\frac{10+17}{21}=\frac{27}{21}$ or $1\frac{2}{7}$

37. $\frac{5}{6}+\frac{5}{6}=\frac{5+5}{6}=\frac{10}{6}$ or $1\frac{2}{3}$

38. $\frac{1}{2}-\frac{1}{2}=\frac{1-1}{2}=\frac{0}{2}$ or $0$

39. $\frac{27}{30}-\frac{12}{30}=\frac{27-12}{30}=\frac{15}{30}$ or $\frac{1}{2}$

40. $\frac{16}{20}-\frac{12}{20}=\frac{16-12}{20}=\frac{4}{20}$ or $\frac{1}{5}$

41. $\frac{21}{24}+\frac{9}{24}=\frac{21+9}{24}=\frac{30}{24}$ or $1\frac{1}{4}$

42. $\frac{15}{25}+\frac{20}{25}=\frac{15+20}{25}=\frac{35}{25}$ or $1\frac{2}{5}$

43. $\frac{8}{12}+\frac{10}{12}=\frac{8+10}{12}=\frac{18}{12}=1\frac{1}{2}$

44. $\frac{10}{16}+\frac{14}{16}=\frac{10+14}{16}=\frac{28}{16}=1\frac{3}{4}$

## Equivalent Fractions (page 6)

1. $\frac{1}{2}=\frac{4}{8}$  2. $\frac{2}{3}=\frac{4}{6}$  3. $\frac{5}{8}=\frac{10}{16}$  4. 4  5. 3  6. 5  7. 3  8. 2  9. 4  10. 12  11. 8  12. $\frac{4}{5}=\frac{20}{25}$

13. $\frac{5}{8}=\frac{15}{24}$  14. $\frac{3}{4}=\frac{12}{16}$  15. $\frac{9}{10}=\frac{27}{30}$  16. $\frac{10}{12}=\frac{40}{48}$  17. $\frac{2}{14}=\frac{8}{56}$  18. $\frac{2}{7}=\frac{14}{49}$  19. $\frac{3}{16}=\frac{27}{144}$

20. $\frac{5}{7}=\frac{30}{42}$  21. $\frac{6}{9}=\frac{12}{18}$  22. $\frac{1}{13}=\frac{3}{39}$  23. $\frac{4}{13}=\frac{44}{143}$  24. $\frac{1}{2}=\frac{8}{16}$  25. $\frac{5}{6}=\frac{10}{12}$  26. $\frac{2}{3}=\frac{8}{12}$

27. $\frac{2}{7}=\frac{4}{14}$  28. $\frac{1}{3}=\frac{9}{27}$  29. $\frac{5}{8}=\frac{15}{24}$  30. $\frac{2}{3}=\frac{6}{9}$  31. $\frac{3}{5}=\frac{12}{20}$  32. $\frac{3}{8}=\frac{12}{32}$  33. $\frac{9}{10}=\frac{45}{50}$

34. $\frac{2}{5}=\frac{12}{30}$  35. $\frac{1}{4}=\frac{13}{52}$

## Least Common Multiple (pages 7–8)

1. 6, 12, 18, 24, 30  2. 5, 10, 15, 20, 25  3. 4, 8, 12, 16, 20  4. 3, 6, 9, 12, 15  5. 15, 30, 45, 60, 75

6. 9, 18, 27, 36, 45  7. 14, 28, 42, 56, 70  8. 10, 20, 30, 40, 50  9. 7, 14, 21, 28, 35

10. 3, 6, 9, 12
    2, 4, 6, 8
    LCM = 6

11. 3, 6, 9, 12
    6, 12, 18, 24
    LCM = 6

12. 6, 12, 18, 24
    8, 16, 24, 32
    LCM = 24

13. 4, 8, 12, 16
    6, 12, 18, 24
    LCM = 12

14. 5, 10, 15, 20, 25, 30, 35, 40
    8, 16, 24, 32, 40
    LCM = 40

15. 6, 12, 18, 24, 30
    15, 30, 45, 60
    LCM = 30

16. 7, 14, 21, 28, 35, 42, 49, 56, 63
    9, 18, 27, 36, 45, 54, 63
    LCM = 63

17. 6, 12, 18, 24, 30
    10, 20, 30, 40
    LCM = 30

18. 8, 16, 24, 32, 40, 48, 56, 64, 72
    18, 36, 54, 72
    LCM = 72

19. 7, 14, 21, 28, 35, 42, 49, 56, 63, 70, 77, 84
    12, 24, 36, 48, 60, 72, 84
    LCM = 84

20. $\frac{20}{28}<\frac{21}{28}$  21. $\frac{27}{72}<\frac{32}{72}$  22. $\frac{9}{42}>\frac{7}{42}$  23. $\frac{16}{20}>\frac{15}{20}$  24. $\frac{5}{10}<\frac{6}{10}$  25. $\frac{20}{24}<\frac{21}{24}$  26. $\frac{21}{60}<\frac{24}{60}$

27. $\frac{16}{18}>\frac{15}{18}$  28. $\frac{81}{90}>\frac{75}{90}$  29. $\frac{8}{12}<\frac{9}{12}$  30. composite  31. composite  32. prime  33. composite

34. prime  35. composite  36. composite  37. composite  38. composite  39. prime  40. prime

41. composite  42. 3, 3  43. 3, 10; 3, 2, 5  44. 2, 2  45. 25; 5, 5

46. 8
    $2\times4$
    $2\times2\times2$

47. 9
    $3\times3$

48. 14
    $2\times7$

49. 18
    $2\times9$
    $2\times3\times3$

50. 20
    $2\times10$
    $2\times2\times5$

51. 16
    $2\times8$
    $2\times2\times4$
    $2\times2\times2\times2$

52. $6=2\times3$
    $10=2\times5$
    LCM $=2\times3\times5$
    LCM = 30

53. $6=2\times3$
    $15=3\times5$
    LCM $=2\times3\times5$
    LCM = 30

54. $7=7$
    $9=3\times3$
    LCM $=3\times3\times7$
    LCM = 63

55. $12=2\times2\times3$
    $18=2\times3\times3$
    LCM $=2\times2\times3\times3$
    LCM = 36

56. $8=2\times2\times2$
    $12=2\times2\times3$
    LCM $=2\times2\times2\times3$
    LCM = 24

57. $15=3\times5$
    $20=2\times2\times5$
    LCM $=2\times2\times3\times5$
    LCM = 60

58. $8=2\times2\times2$
    $18=2\times3\times3$
    LCM $=2\times2\times2\times3\times3$
    LCM = 72

59. $10=2\times5$
    $18=2\times3\times3$
    LCM $=2\times3\times3\times5$
    LCM = 90

60. $2=2$
    $5=5$
    LCM $=2\times5$
    LCM = 10

61. $12=2\times2\times3$
    $35=5\times7$
    LCM $=2\times2\times3\times5\times7$
    LCM = 420

62. $7=7$
    $12=2\times2\times3$
    LCM $=2\times2\times3\times7$
    LCM = 84

63. $14=2\times7$
    $10=2\times5$
    LCM $=2\times5\times7$
    LCM = 70

64. $9=3\times3$
    $12=2\times2\times3$
    $15=3\times5$
    LCM $=2\times2\times3\times3\times5$
    LCM = 180

65. $3=3$
    $6=2\times3$
    $9=3\times3$
    LCM $=2\times3\times3$
    LCM = 18

66. $4=2\times2$
    $8=2\times2\times2$
    $12=2\times2\times3$
    LCM $=2\times2\times2\times3$
    LCM = 24

67. $3=3$
    $7=7$
    LCM $=3\times7$
    LCM = 21

68. $4=2\times2$
    $5=5$
    LCM $=2\times2\times5$
    LCM = 20

69. $9=3\times3$
    $5=5$
    LCM $=3\times3\times5$
    LCM = 45

**70.** $12 = 2 \times 2 \times \underline{3}$
$8 = \underline{2 \times 2 \times 2}$
$LCM = 2 \times 2 \times 2 \times 3$
$LCM = 24$

**71.** $8 = \underline{2 \times 2 \times 2}$
$6 = 2 \times \underline{3}$
$LCM = 2 \times 2 \times 2 \times 3$
$LCM = 24$

**72.** $4 = \underline{2 \times 2}$
$7 = \underline{7}$
$LCM = 2 \times 2 \times 7$
$LCM = 28$

**73.** $9 = \underline{3 \times 3}$
$4 = \underline{2 \times 2}$
$LCM = 2 \times 2 \times 3 \times 3$
$LCM = 36$

**74.** $12 = \underline{2 \times 2} \times 3$
$10 = 2 \times \underline{5}$
$LCM = 2 \times 2 \times 3 \times 5$
$LCM = 60$

**75.** $20 = \underline{2 \times 2 \times 5}$
$15 = \underline{3} \times 5$
$LCM = 2 \times 2 \times 3 \times 5$
$LCM = 60$

**76.** $15 = \underline{3 \times 5}$
$6 = \underline{2} \times 3$
$LCM = 2 \times 3 \times 5$
$LCM = 30$

**77.** $\frac{4}{10} < \frac{7}{10}$

**78.** $\frac{6}{20} = \frac{6}{20}$

**79.** $\frac{14}{21} > \frac{9}{21}$

**80.** $\frac{27}{36} > \frac{20}{36}$

**81.** $\frac{30}{140} < \frac{49}{140}$

## Addition and Subtraction of Fractions (pages 9–10)

**1.**
$\frac{1}{4} = \frac{5}{20}$
$-\frac{1}{5} = \frac{4}{20}$
$\frac{1}{20}$

**2.**
$\frac{1}{2} = \frac{3}{6}$
$-\frac{1}{3} = \frac{2}{6}$
$\frac{1}{6}$

**3.**
$\frac{3}{5} = \frac{9}{15}$
$-\frac{1}{3} = \frac{5}{15}$
$\frac{4}{15}$

**4.**
$\frac{7}{8} = \frac{21}{24}$
$-\frac{1}{6} = \frac{4}{24}$
$\frac{17}{24}$

**5.**
$\frac{1}{4} = \frac{2}{8}$
$+\frac{1}{8} = \frac{1}{8}$
$\frac{3}{8}$

**6.**
$\frac{1}{3} = \frac{2}{6}$
$+\frac{5}{6} = \frac{5}{6}$
$\frac{7}{6}$
or $1\frac{1}{6}$

**7.**
$\frac{2}{9} = \frac{2}{9}$
$+\frac{2}{3} = \frac{6}{9}$
$\frac{8}{9}$

**8.**
$\frac{3}{4} = \frac{6}{8}$
$+\frac{5}{8} = \frac{5}{8}$
$\frac{11}{8}$
or $1\frac{3}{8}$

**9.**
$\frac{3}{4} = \frac{9}{12}$
$-\frac{2}{3} = \frac{8}{12}$
$\frac{1}{12}$

**10.**
$\frac{2}{3} = \frac{8}{12}$
$-\frac{1}{4} = \frac{3}{12}$
$\frac{5}{12}$

**11.**
$\frac{6}{7} = \frac{18}{21}$
$-\frac{2}{3} = \frac{14}{21}$
$\frac{4}{21}$

**12.**
$\frac{5}{6} = \frac{15}{18}$
$\frac{2}{9} = \frac{4}{18}$
$\frac{11}{18}$

**13.**
$\frac{1}{8} = \frac{3}{24}$
$+\frac{5}{6} = \frac{20}{24}$
$\frac{23}{24}$

**14.**
$\frac{2}{5} = \frac{14}{35}$
$\frac{3}{7} = \frac{15}{35}$
$\frac{29}{35}$

**15.**
$\frac{3}{4} = \frac{15}{20}$
$+\frac{2}{5} = \frac{8}{20}$
$\frac{23}{20}$
or $1\frac{3}{20}$

**16.**
$\frac{1}{3} = \frac{8}{24}$
$+\frac{3}{8} = \frac{9}{24}$
$\frac{17}{24}$

**17.**
$\frac{3}{20} = \frac{15}{100}$
$+\frac{3}{25} = \frac{12}{100}$
$\frac{27}{100}$
of the population of Oklahoma

**18.**
$\frac{3}{10} = \frac{6}{20}$
$+\frac{3}{20} = \frac{3}{20}$
$\frac{9}{20}$
of the population of Arizona

**19.**
$\frac{5}{6} = \frac{10}{12}$
$+\frac{1}{4} = \frac{3}{12}$
$\frac{13}{12}$
or $1\frac{1}{12}$

**20.**
$\frac{3}{8} = \frac{9}{24}$
$+\frac{5}{6} = \frac{20}{24}$
$\frac{29}{24}$
or $1\frac{5}{24}$

**21.**
$\frac{5}{6} = \frac{15}{18}$
$+\frac{2}{9} = \frac{4}{18}$
$\frac{19}{18}$
or $1\frac{1}{18}$

**22.**
$\frac{6}{7} = \frac{18}{21}$
$-\frac{1}{3} = \frac{7}{21}$
$\frac{11}{21}$

**23.**
$\frac{7}{9} = \frac{14}{18}$
$-\frac{1}{2} = \frac{9}{18}$
$\frac{5}{18}$

**24.**
$\frac{3}{7} = \frac{15}{35}$
$-\frac{2}{5} = \frac{14}{35}$
$\frac{1}{35}$

**25.**
$\frac{8}{15} = \frac{8}{15}$
$+\frac{1}{5} = \frac{3}{15}$
$\frac{11}{15}$

**26.**
$\frac{7}{8} = \frac{21}{24}$
$-\frac{5}{6} = \frac{20}{24}$
$\frac{1}{24}$

**27.**
$\frac{1}{5} = \frac{6}{30}$
$+\frac{5}{6} = \frac{25}{30}$
$\frac{31}{30}$
or $1\frac{1}{30}$

**28.**
$\frac{5}{8} = \frac{15}{24}$
$+\frac{5}{6} = \frac{20}{24}$
$\frac{35}{24}$
or $1\frac{11}{24}$

**29.**
$\frac{11}{12} = \frac{11}{12}$
$+\frac{2}{3} = \frac{8}{12}$
$\frac{19}{12}$
or $1\frac{7}{12}$

**30.**
$\frac{6}{7} = \frac{18}{21}$
$-\frac{1}{3} = \frac{7}{21}$
$\frac{11}{21}$

**31.**
$\frac{13}{24} = \frac{13}{24}$
$-\frac{1}{8} = \frac{3}{24}$
$\frac{10}{24}$
or $\frac{5}{12}$

**32.**
$\frac{3}{7} = \frac{9}{21}$
$-\frac{5}{21} = \frac{5}{21}$
$\frac{4}{21}$

**33.**
$\frac{14}{15} = \frac{14}{15}$
$-\frac{1}{3} = \frac{5}{15}$
$\frac{9}{15}$
or $\frac{3}{5}$

**34.**
$\frac{11}{12} = \frac{11}{12}$
$-\frac{2}{3} = \frac{8}{12}$
$\frac{3}{12}$
or $\frac{1}{4}$

**35.**
$\frac{2}{3} = \frac{14}{21}$
$+\frac{3}{7} = \frac{9}{21}$
$\frac{23}{21}$
or $1\frac{2}{21}$

**36.**
$\frac{7}{9} = \frac{56}{72}$
$+\frac{3}{8} = \frac{27}{72}$
$\frac{83}{72}$
or $1\frac{11}{72}$

**37.**
$\frac{7}{20} = \frac{7}{20}$
$+\frac{4}{5} = \frac{16}{20}$
$\frac{23}{20}$
or $1\frac{3}{20}$

**38.**
$\frac{4}{5} = \frac{16}{20}$
$-\frac{11}{20} = \frac{11}{20}$
$\frac{5}{20}$
or $\frac{1}{4}$

**39.**
$\frac{9}{10} = \frac{81}{90}$
$-\frac{2}{9} = \frac{20}{90}$
$\frac{61}{90}$

**40.**
$\frac{11}{12} = \frac{55}{60}$
$-\frac{2}{5} = \frac{24}{60}$
$\frac{31}{60}$

29

**41.** $\frac{3}{7}=\frac{15}{35}$
$+\frac{2}{5}=\frac{14}{35}$
$\frac{29}{35}$

**42.** $\frac{2}{9}=\frac{4}{18}$
$+\frac{1}{6}=\frac{3}{18}$
$\frac{7}{18}$

**43.** $\frac{5}{9}=\frac{25}{45}$
$+\frac{2}{5}=\frac{18}{45}$
$\frac{43}{45}$

**44.** $\frac{3}{4}=\frac{15}{20}$
$+\frac{3}{10}=\frac{6}{20}$
$\frac{21}{20}$
or $1\frac{1}{20}$

**45.** $\frac{6}{7}=\frac{54}{63}$
$-\frac{2}{9}=\frac{14}{63}$
$\frac{40}{63}$

**46.** $\frac{14}{15}=\frac{28}{30}$
$-\frac{5}{6}=\frac{25}{30}$
$\frac{3}{30}$
or $\frac{1}{10}$

**47.** $\frac{5}{6}=\frac{10}{12}$
$-\frac{1}{4}=\frac{3}{12}$
$\frac{7}{12}$

**48.** $\frac{7}{8}=\frac{21}{24}$
$-\frac{5}{12}=\frac{14}{24}$
$\frac{7}{24}$

**49.** $\frac{9}{10}=\frac{54}{60}$
$-\frac{5}{12}=\frac{25}{60}$
$\frac{29}{60}$

**50.** $\frac{5}{6}=\frac{20}{24}$
$-\frac{3}{8}=\frac{9}{24}$
$\frac{11}{24}$

**51.** $\frac{1}{3}=\frac{5}{15}$
$+\frac{3}{5}=\frac{9}{15}$
$\frac{14}{15}$

**52.** $\frac{5}{6}=\frac{25}{30}$
$+\frac{2}{5}=\frac{12}{30}$
$\frac{37}{30}$
or $1\frac{7}{30}$

**53.** $\frac{5}{9}=\frac{40}{72}$
$+\frac{7}{8}=\frac{63}{72}$
$\frac{103}{72}$
or $1\frac{31}{72}$

**54.** $\frac{1}{3}=\frac{7}{21}$
$+\frac{3}{7}=\frac{9}{21}$
$\frac{16}{21}$

**55.** $\frac{1}{4}=\frac{5}{20}$
$+\frac{1}{20}=\frac{1}{20}$
$\frac{6}{20}$
or $\frac{3}{10}$
of transportation energy is used by trucks and airplanes

**56.** $\frac{1}{2}=\frac{3}{6}$
$+\frac{1}{6}=\frac{1}{6}$
$\frac{4}{6}$
or $\frac{2}{3}$
of transportation energy is used by car, buses, and railroads

## Addition and Subtraction of Mixed Numerals (pages 11–12)

**1.** $1\frac{1}{4}$ $+3\frac{2}{4}$ $4\frac{3}{4}$

**2.** $8\frac{3}{8}$ $+3\frac{1}{8}$ $11\frac{4}{8}$ or $11\frac{1}{2}$

**3.** $13\frac{2}{5}$ $+7\frac{2}{5}$ $20\frac{4}{5}$

**4.** $6\frac{3}{12}$ $+3\frac{5}{12}$ $9\frac{8}{12}$ or $9\frac{2}{3}$

**5.** $4\frac{1}{3}$ $+2\frac{2}{3}$ $6\frac{3}{3}$ or $7$

**6.** $2\frac{7}{10}$ $+9\frac{9}{10}$ $11\frac{16}{10}$ $=12\frac{6}{10}$ or $12\frac{3}{5}$

**7.** $24\frac{5}{7}$ $+15\frac{4}{7}$ $39\frac{9}{7}$ or $40\frac{2}{7}$

**8.** $21\frac{1}{6}$ $+7\frac{5}{6}$ $28\frac{6}{6}$ or $29$

**9.** $8\frac{9}{16}$ $+4\frac{11}{16}$ $12\frac{20}{16}$ $=13\frac{4}{16}$ or $13\frac{1}{4}$

**10.** $15\frac{7}{8}$ $+4\frac{5}{8}$ $19\frac{12}{8}$ $=20\frac{4}{8}$ or $20\frac{1}{2}$

**11.** $7\frac{7}{9}$ $-1\frac{3}{9}$ $6\frac{4}{9}$

**12.** $9\frac{3}{4}$ $-2\frac{1}{4}$ $7\frac{2}{4}$ or $7\frac{1}{2}$

**13.** $8\frac{4}{5}$ $-2\frac{1}{5}$ $6\frac{3}{5}$

**14.** $17\frac{9}{10}$ $-6\frac{3}{10}$ $11\frac{6}{10}$ or $11\frac{3}{5}$

**15.** $29\frac{5}{7}$ $-18\frac{2}{7}$ $11\frac{3}{7}$

**16.** $13\frac{7}{8}$ $-2\frac{3}{8}$ $11\frac{4}{8}$ or $11\frac{1}{2}$

**17.** $8\frac{3}{5}=8\frac{12}{20}$ $-5\frac{1}{4}=5\frac{5}{20}$ $3\frac{7}{20}$

**18.** $7\frac{2}{3}=7\frac{10}{15}$ $-2\frac{1}{5}=2\frac{3}{15}$ $5\frac{7}{15}$

**19.** $12\frac{11}{12}=12\frac{11}{12}$ $-5\frac{1}{4}=5\frac{3}{12}$ $7\frac{8}{12}$ or $7\frac{2}{3}$

**20.** $37\frac{5}{6}=37\frac{20}{24}$ $-11\frac{3}{8}=11\frac{9}{24}$ $26\frac{11}{24}$

**21.** $40\frac{4}{5}$ inches of rain in Montreal
$-31\frac{3}{5}$ inches of rain in Seattle
$9\frac{1}{5}$ inches of rain difference

**22.** $6\frac{1}{2}=6\frac{5}{10}$ thousand tons of tin mined by Bolivia
$+3\frac{9}{10}=3\frac{9}{10}$ thousand tons of tin mined by Belgium
$9\frac{14}{10}$ thousand tons of tin mixed together
$=10\frac{4}{10}$ or $10\frac{2}{5}$

**23.** $38\frac{3}{7}=38\frac{30}{70}=37\frac{100}{70}$ $-29\frac{9}{10}=29\frac{63}{70}=29\frac{63}{70}$ $8\frac{37}{70}$

**24.** $6\frac{2}{3}=6\frac{6}{9}$ $-1\frac{5}{9}=1\frac{5}{9}$ $5\frac{1}{9}$

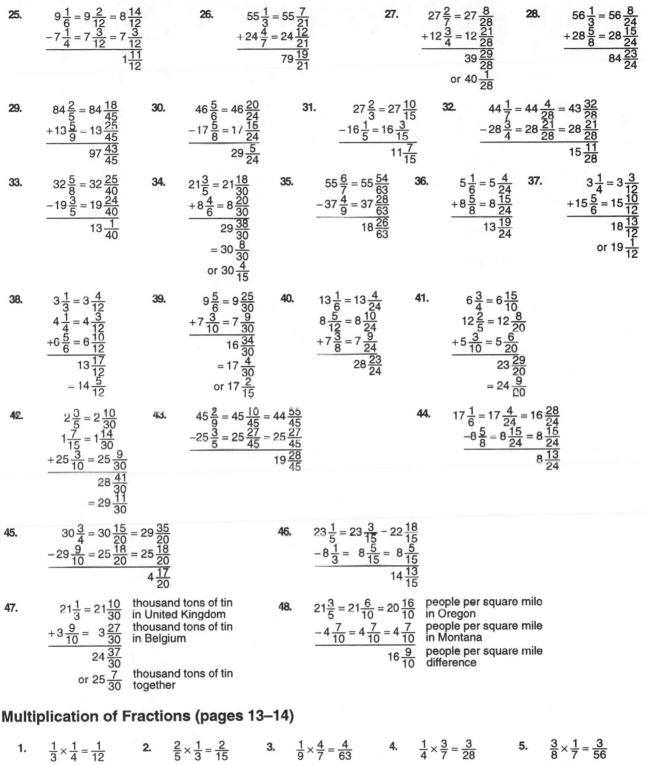

**25.**
$$9\frac{1}{6} = 9\frac{2}{12} = 8\frac{14}{12}$$
$$-7\frac{1}{4} = 7\frac{3}{12} = 7\frac{3}{12}$$
$$1\frac{11}{12}$$

**26.**
$$55\frac{1}{3} = 55\frac{7}{21}$$
$$+24\frac{4}{7} = 24\frac{12}{21}$$
$$79\frac{19}{21}$$

**27.**
$$27\frac{2}{7} = 27\frac{8}{28}$$
$$+12\frac{3}{4} = 12\frac{21}{28}$$
$$39\frac{29}{28}$$
$$\text{or } 40\frac{1}{28}$$

**28.**
$$56\frac{1}{3} = 56\frac{8}{24}$$
$$+28\frac{5}{8} = 28\frac{15}{24}$$
$$84\frac{23}{24}$$

**29.**
$$84\frac{2}{5} = 84\frac{18}{45}$$
$$+13\frac{5}{9} = 13\frac{25}{45}$$
$$97\frac{43}{45}$$

**30.**
$$46\frac{5}{6} = 46\frac{20}{24}$$
$$-17\frac{5}{8} = 17\frac{16}{24}$$
$$29\frac{5}{24}$$

**31.**
$$27\frac{2}{3} = 27\frac{10}{15}$$
$$-16\frac{1}{5} = 16\frac{3}{15}$$
$$11\frac{7}{15}$$

**32.**
$$44\frac{1}{7} = 44\frac{4}{28} = 43\frac{32}{28}$$
$$-28\frac{3}{4} = 28\frac{21}{28} = 28\frac{21}{28}$$
$$15\frac{11}{28}$$

**33.**
$$32\frac{5}{8} = 32\frac{25}{40}$$
$$-19\frac{3}{5} = 19\frac{24}{40}$$
$$13\frac{1}{40}$$

**34.**
$$21\frac{3}{5} = 21\frac{18}{30}$$
$$+8\frac{4}{6} = 8\frac{20}{30}$$
$$29\frac{38}{30}$$
$$= 30\frac{8}{30}$$
$$\text{or } 30\frac{4}{15}$$

**35.**
$$55\frac{6}{7} = 55\frac{54}{63}$$
$$-37\frac{4}{9} = 37\frac{28}{63}$$
$$18\frac{26}{63}$$

**36.**
$$5\frac{1}{6} = 5\frac{4}{24}$$
$$+8\frac{5}{8} = 8\frac{15}{24}$$
$$13\frac{19}{24}$$

**37.**
$$3\frac{1}{4} = 3\frac{3}{12}$$
$$+15\frac{5}{6} = 15\frac{10}{12}$$
$$18\frac{13}{12}$$
$$\text{or } 19\frac{1}{12}$$

**38.**
$$3\frac{1}{3} = 3\frac{4}{12}$$
$$4\frac{1}{4} = 4\frac{3}{12}$$
$$+6\frac{5}{6} = 6\frac{10}{12}$$
$$13\frac{17}{12}$$
$$= 14\frac{5}{12}$$

**39.**
$$9\frac{5}{6} = 9\frac{25}{30}$$
$$+7\frac{3}{10} = 7\frac{9}{30}$$
$$16\frac{34}{30}$$
$$= 17\frac{4}{30}$$
$$\text{or } 17\frac{2}{15}$$

**40.**
$$13\frac{1}{6} = 13\frac{4}{24}$$
$$8\frac{5}{12} = 8\frac{10}{24}$$
$$+7\frac{3}{8} = 7\frac{9}{24}$$
$$28\frac{23}{24}$$

**41.**
$$6\frac{3}{4} = 6\frac{15}{10}$$
$$12\frac{2}{5} = 12\frac{8}{20}$$
$$+5\frac{3}{10} = 5\frac{6}{20}$$
$$23\frac{29}{20}$$
$$= 24\frac{9}{20}$$

**42.**
$$2\frac{9}{5} = 2\frac{10}{30}$$
$$1\frac{7}{15} = 1\frac{14}{30}$$
$$+25\frac{3}{10} = 25\frac{9}{30}$$
$$28\frac{41}{30}$$
$$= 29\frac{11}{30}$$

**43.**
$$45\frac{2}{9} = 45\frac{10}{45} = 44\frac{55}{45}$$
$$-25\frac{3}{5} = 25\frac{27}{45} = 25\frac{27}{45}$$
$$19\frac{28}{45}$$

**44.**
$$17\frac{1}{6} = 17\frac{4}{24} = 16\frac{28}{24}$$
$$-8\frac{5}{8} = 8\frac{15}{24} = 8\frac{15}{24}$$
$$8\frac{13}{24}$$

**45.**
$$30\frac{3}{4} = 30\frac{15}{20} = 29\frac{35}{20}$$
$$-29\frac{9}{10} = 25\frac{18}{20} = 25\frac{18}{20}$$
$$4\frac{17}{20}$$

**46.**
$$23\frac{1}{5} = 23\frac{3}{15} - 22\frac{18}{15}$$
$$-8\frac{1}{3} = 8\frac{5}{15} = 8\frac{5}{15}$$
$$14\frac{13}{15}$$

**47.**
$$21\frac{1}{3} = 21\frac{10}{30} \quad \text{thousand tons of tin in United Kingdom}$$
$$+3\frac{9}{10} = 3\frac{27}{30} \quad \text{thousand tons of tin in Belgium}$$
$$24\frac{37}{30}$$
$$\text{or } 25\frac{7}{30} \quad \text{thousand tons of tin together}$$

**48.**
$$21\frac{3}{5} = 21\frac{6}{10} = 20\frac{16}{10} \quad \text{people per square mile in Oregon}$$
$$-4\frac{7}{10} = 4\frac{7}{10} = 4\frac{7}{10} \quad \text{people per square mile in Montana}$$
$$16\frac{9}{10} \quad \text{people per square mile difference}$$

## Multiplication of Fractions (pages 13–14)

**1.** $\frac{1}{3} \times \frac{1}{4} = \frac{1}{12}$

**2.** $\frac{2}{5} \times \frac{1}{3} = \frac{2}{15}$

**3.** $\frac{1}{9} \times \frac{4}{7} = \frac{4}{63}$

**4.** $\frac{1}{4} \times \frac{3}{7} = \frac{3}{28}$

**5.** $\frac{3}{8} \times \frac{1}{7} = \frac{3}{56}$

**6.** $\frac{5}{8} \times \frac{3}{4} = \frac{15}{32}$

**7.** $\frac{2}{7} \times \frac{7}{12} = \frac{14}{84} \text{ or } \frac{1}{6}$

**8.** $\frac{4}{7} \times \frac{3}{10} = \frac{12}{70} \text{ or } \frac{6}{35}$

**9.** $\frac{6}{7} \times \frac{14}{15} = \frac{\overset{2}{6}}{7} \times \frac{\overset{2}{14}}{15} = \frac{4}{5}$

**10.** $\frac{8}{15} \times \frac{5}{6} = \frac{\overset{4}{8}}{\underset{3}{15}} \times \frac{\overset{1}{5}}{\underset{3}{6}} = \frac{4}{9}$

**11.** $\frac{5}{6} \times \frac{3}{8} = \frac{5}{\underset{2}{6}} \times \frac{\overset{1}{3}}{8} = \frac{5}{16}$

**12.** $\frac{3}{8} \times \frac{4}{9} = \frac{\overset{1}{3}}{\underset{2}{8}} \times \frac{\overset{1}{4}}{\underset{3}{9}} = \frac{1}{6}$

**13.** $\frac{7}{8} \times \frac{4}{7} = \frac{\overset{1}{7}}{\underset{2}{8}} \times \frac{\overset{1}{4}}{\underset{1}{7}} = \frac{1}{2}$

31

14. $\frac{7}{16} \times \frac{8}{21} = \frac{\overset{1}{7}}{\underset{2}{16}} \times \frac{\overset{1}{8}}{\underset{3}{21}} = \frac{1}{6}$

15. $\frac{5}{9} \times \frac{3}{5} = \frac{\overset{1}{5}}{\underset{3}{9}} \times \frac{\overset{1}{3}}{\underset{1}{5}} = \frac{1}{3}$

16. $\frac{2}{3} \times \frac{1}{2} \times \frac{3}{4} = \frac{\overset{1}{2}}{\underset{1}{3}} \times \frac{1}{\underset{1}{2}} \times \frac{\overset{1}{3}}{4} = \frac{1}{4}$

17. $\frac{3}{16} \times \frac{4}{9} = \frac{3}{\underset{4}{16}} \times \frac{\overset{1}{4}}{\underset{3}{9}} = \frac{1}{12}$

18. $\frac{3}{8} \times \frac{2}{3} \times \frac{1}{4} = \frac{3}{\underset{4}{8}} \times \frac{\overset{1}{2}}{\underset{1}{3}} \times \frac{1}{4} = \frac{1}{16}$

19. $\frac{3}{7} \times \frac{11}{15} \times \frac{7}{12} = \frac{3}{7} \times \frac{11}{\underset{5}{15}} \times \frac{\overset{1}{7}}{12} = \frac{11}{60}$

20. $\frac{2}{3} \times \frac{3}{5} \times \frac{6}{7} = \frac{2}{3} \times \frac{3}{5} \times \frac{\overset{1}{6}}{7} = \frac{12}{35}$

21. $\frac{3}{4} \times \frac{5}{8} \times \frac{4}{5} = \frac{3}{\underset{1}{4}} \times \frac{\overset{1}{5}}{8} \times \frac{\overset{1}{4}}{\underset{1}{5}} = \frac{3}{8}$

22. $\frac{4}{5} \times \frac{1}{3} \times \frac{15}{16} = \frac{\overset{1}{4}}{5} \times \frac{1}{\underset{1}{3}} \times \frac{\overset{1}{15}}{\underset{4}{16}} = \frac{1}{4}$

23. $\frac{8}{9} \times \frac{3}{5} \times \frac{15}{16} = \frac{\overset{1}{8}}{\underset{3}{9}} \times \frac{\overset{1}{3}}{\underset{1}{5}} \times \frac{\overset{3}{15}}{\underset{2}{16}} = \frac{1}{2}$

24. $\frac{9}{16} \times \frac{2}{3} \times \frac{5}{6} = \frac{\overset{3}{9}}{\underset{8}{16}} \times \frac{\overset{1}{2}}{3} \times \frac{5}{\underset{2}{6}} = \frac{5}{16}$

25. $5 \times \frac{1}{8} = \frac{5}{1} \times \frac{1}{8} = \frac{5}{8}$

26. $\frac{1}{3} \times 4 = \frac{1}{3} \times \frac{4}{1} = \frac{4}{3}$ or $1\frac{1}{3}$

27. $8 \times \frac{2}{7} = \frac{8}{1} \times \frac{2}{7} = \frac{16}{7}$ or $2\frac{2}{7}$

28. $\frac{7}{9} \times 3 = \frac{7}{9} \times \frac{3}{1} = \frac{21}{9}$ or $2\frac{1}{3}$

29. $\frac{3}{4} \times 7 = \frac{3}{4} \times \frac{7}{1} = \frac{21}{4}$ or $5\frac{1}{4}$

30. $\frac{1}{5} \times 13 = \frac{1}{5} \times \frac{13}{1} = \frac{13}{5}$ or $2\frac{3}{5}$

31. $5 \times \frac{5}{8} = \frac{5}{1} \times \frac{5}{8} = \frac{25}{8}$ or $3\frac{1}{8}$

32. $5 \times \frac{1}{9} = \frac{5}{1} \times \frac{1}{9} = \frac{5}{9}$

33. $\frac{5}{8} \times 4 = \frac{5}{\underset{2}{8}} \times \frac{\overset{1}{4}}{1} = \frac{5}{2}$ or $2\frac{1}{2}$

34. $6 \times \frac{1}{3} = \frac{\overset{2}{6}}{1} \times \frac{1}{\underset{1}{3}} = 2$

35. $\frac{3}{8} \times 14 = \frac{3}{\underset{4}{8}} \times \frac{\overset{7}{14}}{1} = \frac{21}{4}$ or $5\frac{1}{4}$

36. $12 \times \frac{2}{9} = \frac{\overset{4}{12}}{1} \times \frac{2}{\underset{3}{9}} = \frac{8}{3}$ or $2\frac{2}{3}$

37. $\frac{3}{4} \times 8 = \frac{3}{\underset{1}{4}} \times \frac{\overset{2}{8}}{1} = 6$

38. $3 \times \frac{5}{9} = \frac{\overset{1}{3}}{1} \times \frac{5}{\underset{3}{9}} = \frac{5}{3}$ or $1\frac{2}{3}$

39. $8 \times \frac{9}{10} = \frac{\overset{4}{8}}{1} \times \frac{9}{\underset{5}{10}} = \frac{36}{5}$ or $7\frac{1}{5}$

40. $6 \times \frac{1}{4} = \frac{\overset{3}{6}}{1} \times \frac{1}{\underset{2}{4}} = \frac{3}{2}$ or $1\frac{1}{2}$

41. $\frac{7}{15} \times 20 = \frac{7}{\underset{3}{15}} \times \frac{\overset{4}{20}}{1} = \frac{28}{3}$ or $9\frac{1}{3}$

42. $10 \times \frac{3}{14} = \frac{\overset{5}{10}}{1} \times \frac{3}{\underset{7}{14}} = \frac{15}{7}$ or $2\frac{1}{7}$

43. $\frac{1}{6} \times \frac{2}{5} \times 20 = \frac{1}{\underset{3}{6}} \times \frac{\overset{1}{2}}{\underset{1}{5}} \times \frac{\overset{4}{20}}{1} = \frac{4}{3}$ or $1\frac{1}{3}$

44. $\frac{5}{12} \times 18 = \frac{5}{\underset{2}{12}} \times \frac{\overset{3}{18}}{1} = \frac{15}{2} = 7\frac{1}{2}$

45. $\frac{2}{3} \times 9 \times \frac{5}{6} \times 9 = \frac{2}{\underset{1}{3}} \times \overset{3}{9} \times \frac{5}{\underset{2}{6}} = 5$

46. $\frac{7}{8} \times 16 = \frac{7}{\underset{1}{8}} \times \frac{\overset{2}{16}}{1} = 14$

47. $3 \times 4 \times \frac{5}{6} = \frac{\overset{1}{3}}{1} \times \frac{\overset{2}{4}}{1} \times \frac{5}{\underset{2}{6}} = 10$

48. $5 \times \frac{2}{15} \times \frac{3}{4} = \frac{\overset{1}{5}}{1} \times \frac{\overset{1}{2}}{\underset{3}{15}} \times \frac{\overset{1}{3}}{\underset{2}{4}} = \frac{1}{2}$

49. $\frac{7}{10} \times \frac{5}{7} = \frac{\overset{1}{7}}{\underset{2}{10}} \times \frac{\overset{1}{5}}{\underset{1}{7}} = \frac{1}{2}$ are beans

50. $\frac{3}{4} \times \frac{4}{9} = \frac{\overset{1}{3}}{\underset{1}{4}} \times \frac{\overset{1}{4}}{\underset{3}{9}} = \frac{1}{3}$ are tulips

51. $\frac{2}{5} \times \frac{1}{2} = \frac{\overset{1}{2}}{5} \times \frac{1}{\underset{1}{2}} = \frac{1}{5}$

52. $\frac{1}{2} \times \frac{6}{13} = \frac{1}{\underset{1}{2}} \times \frac{\overset{3}{6}}{13} = \frac{3}{13}$

53. $\frac{9}{15} \times \frac{10}{27} = \frac{9}{\underset{3}{15}} \times \frac{10}{\underset{3}{27}} = \frac{2}{9}$

54. $\frac{6}{7} \times \frac{7}{10} = \frac{\overset{3}{6}}{\underset{1}{7}} \times \frac{\overset{1}{7}}{\underset{5}{10}} = \frac{3}{5}$

55. $\frac{2}{3} \times \frac{3}{8} = \frac{\overset{1}{2}}{\underset{1}{3}} \times \frac{\overset{1}{3}}{\underset{4}{8}} = \frac{1}{4}$

56. $\frac{2}{25} \times \frac{5}{8} = \frac{\overset{1}{2}}{\underset{5}{25}} \times \frac{\overset{1}{5}}{\underset{4}{8}} = \frac{1}{20}$

57. $\frac{7}{15} \times \frac{5}{7} = \frac{\overset{1}{7}}{\underset{3}{15}} \times \frac{\overset{1}{5}}{\underset{1}{7}} = \frac{1}{3}$

58. $\frac{2}{5} \times \frac{5}{8} = \frac{\overset{1}{2}}{\underset{1}{5}} \times \frac{\overset{1}{5}}{\underset{4}{8}} = \frac{1}{4}$

59. $\frac{4}{5} \times \frac{15}{16} = \frac{\overset{1}{4}}{\underset{1}{5}} \times \frac{\overset{3}{15}}{\underset{4}{16}} = \frac{3}{4}$

60. $\frac{3}{8} \times \frac{2}{7} = \frac{3}{\underset{4}{8}} \times \frac{\overset{1}{2}}{7} = \frac{3}{28}$

61. $\frac{5}{8} \times \frac{4}{15} = \frac{\overset{1}{5}}{\underset{2}{8}} \times \frac{\overset{1}{4}}{\underset{3}{15}} = \frac{1}{6}$

62. $\frac{9}{10} \times \frac{5}{6} = \frac{\overset{3}{9}}{\underset{2}{10}} \times \frac{\overset{1}{5}}{\underset{2}{6}} = \frac{3}{4}$

63. $\frac{5}{6} \times \frac{24}{35} = \frac{\overset{1}{5}}{\underset{1}{6}} \times \frac{\overset{4}{24}}{\underset{7}{35}} = \frac{4}{7}$

64. $\frac{5}{9} \times \frac{3}{20} = \frac{\overset{1}{5}}{\underset{3}{9}} \times \frac{\overset{1}{3}}{\underset{4}{20}} = \frac{1}{12}$

65. $\frac{3}{10} \times \frac{5}{18} = \frac{3}{\underset{2}{10}} \times \frac{\overset{1}{5}}{\underset{6}{18}} = \frac{1}{12}$

66. $\frac{4}{35} \times \frac{5}{16} = \frac{\overset{1}{4}}{\underset{7}{35}} \times \frac{\overset{1}{5}}{\underset{4}{16}} = \frac{1}{28}$

67. $\frac{3}{4} \times \frac{4}{5} \times \frac{5}{12} = \frac{3}{\underset{1}{4}} \times \frac{\overset{1}{4}}{\underset{1}{5}} \times \frac{\overset{1}{5}}{\underset{4}{12}} = \frac{1}{4}$

68. $\frac{1}{4} \times \frac{7}{6} \times \frac{24}{35} = \frac{1}{\cancel{4}} \times \frac{\cancel{7}}{\cancel{6}} \times \frac{\cancel{24}}{\cancel{35}} = \frac{1}{5}$

69. $\frac{2}{5} \times \frac{2}{3} \times \frac{21}{18} = \frac{2}{5} \times \frac{2}{\cancel{3}} \times \frac{\cancel{21}}{\cancel{18}} = \frac{14}{45}$

70. $\frac{1}{5} \times \frac{15}{20} = \frac{1}{5} \times \frac{\cancel{15}}{20} = \frac{3}{20}$

71. $\frac{3}{5} \times \frac{2}{3} \times \frac{45}{50} = \frac{\cancel{3}}{5} \times \frac{2}{\cancel{3}} \times \frac{\cancel{45}}{\cancel{50}} = \frac{9}{25}$

72. $\frac{3}{10} \times \frac{5}{15} \times \frac{20}{25} = \frac{3}{\cancel{10}} \times \frac{\cancel{5}}{\cancel{15}} \times \frac{\cancel{20}}{25} = \frac{2}{25}$

73. $\frac{4}{18} \times \frac{3}{16} = \frac{\cancel{4}}{\cancel{18}} \times \frac{\cancel{3}}{\cancel{16}} = \frac{1}{24}$

74. $\frac{5}{16} \times \frac{4}{15} \times \frac{12}{18} = \frac{\cancel{5}}{\cancel{16}} \times \frac{\cancel{4}}{\cancel{15}} \times \frac{\cancel{12}}{18} = \frac{1}{18}$

## Renaming Mixed Numerals (page 15)

1. $2\frac{1}{3} = \frac{(2\times3)+1}{3} = \frac{7}{3}$

2. $3\frac{1}{2} = \frac{(3\times2)+1}{2} = \frac{7}{2}$

3. $2\frac{3}{5} = \frac{(2\times5)+3}{5} = \frac{13}{5}$

4. $3\frac{5}{8} = \frac{(3\times8)+5}{8} = \frac{29}{8}$

5. $2\frac{4}{5} = \frac{(2\times5)+4}{5} = \frac{14}{5}$

6. $3\frac{1}{6} = \frac{(3\times6)+1}{6} = \frac{19}{6}$

7. $6\frac{3}{7} = \frac{(6\times7)+3}{7} = \frac{45}{7}$

8. $7\frac{1}{3} = \frac{(7\times3)+1}{3} = \frac{22}{3}$

9. $4\frac{3}{4} = \frac{(4\times4)+3}{4} = \frac{19}{4}$

10. $5\frac{2}{3} = \frac{(5\times3)+2}{3} = \frac{17}{3}$

11. $3\frac{9}{10} = \frac{(3\times10)+9}{10} = \frac{39}{10}$

12. $2\frac{7}{12} = \frac{(2\times12)+7}{12} = \frac{31}{12}$

13. $2\frac{5}{8} = \frac{(2\times8)+5}{8} = \frac{21}{8}$

14. $3\frac{2}{3} = \frac{(3\times3)+2}{3} = \frac{11}{3}$

15. $3\frac{5}{6} = \frac{(3\times6)+5}{6} = \frac{23}{6}$

16. $3\frac{4}{9} = \frac{(3\times9)+4}{9} = \frac{31}{9}$

17. $6\frac{5}{8} = \frac{(6\times8)+5}{8} = \frac{53}{8}$

18. $9\frac{2}{3} = \frac{(9\times3)+2}{3} = \frac{29}{3}$

19. $12\frac{1}{2} = \frac{(12\times2)+1}{2} = \frac{25}{2}$

20. $5\frac{7}{16} = \frac{(5\times16)+7}{16} = \frac{87}{16}$

21. $14\frac{2}{3} = \frac{(14\times3)+2}{3} = \frac{44}{3}$

22. $10\frac{3}{10} = \frac{(10\times10)+3}{10} = \frac{103}{10}$

23. $2\frac{13}{16} = \frac{(2\times16)+13}{16} = \frac{45}{16}$

24. $18\frac{4}{7} = \frac{(18\times7)+4}{7} = \frac{130}{7}$

25. $7\frac{3}{8} = \frac{(7\times8)+3}{8} = \frac{59}{8}$

26. $8\frac{5}{9} = \frac{(8\times9)+5}{9} = \frac{77}{9}$

27. $11\frac{3}{10} = \frac{(11\times10)+3}{10} = \frac{113}{10}$

28. $3\frac{5}{12} = \frac{(3\times12)+5}{12} = \frac{41}{12}$

29. $7\frac{4}{15} = \frac{(7\times15)+4}{15} = \frac{109}{15}$

30. $11\frac{4}{5} = \frac{(11\times5)+4}{5} = \frac{59}{5}$

31. $14\frac{2}{5} = \frac{(14\times5)+2}{5} = \frac{72}{5}$

32. $6\frac{1}{15} = \frac{(6\times15)+1}{15} = \frac{91}{15}$

33. $2\frac{11}{12} = \frac{(2\times12)+11}{12} = \frac{35}{12}$

34. $2\frac{3}{4} = \frac{(2\times4)+3}{4} = \frac{11}{4}$

35. $2\frac{1}{9} = \frac{(2\times9)+1}{9} = \frac{19}{9}$

36. $1\frac{3}{5} = \frac{(1\times5)+3}{5} = \frac{8}{5}$

37. $1\frac{6}{7} = \frac{(1\times7)+6}{7} = \frac{13}{7}$

38. $5\frac{7}{12} = \frac{(5\times12)+7}{12} = \frac{67}{12}$

39. $9\frac{3}{8} = \frac{(9\times8)+3}{8} = \frac{75}{8}$

40. $16\frac{1}{3} = \frac{(16\times3)+1}{3} = \frac{49}{3}$

41. $2\frac{8}{9} = \frac{(2\times9)+8}{9} = \frac{26}{9}$

42. $4\frac{1}{10} = \frac{(4\times10)+1}{10} = \frac{41}{10}$

43. $20\frac{5}{6} = \frac{(20\times6)+5}{6} = \frac{125}{6}$

44. $6\frac{6}{7} = \frac{(6\times7)+6}{7} = \frac{48}{7}$

45. $2\frac{1}{3} = \frac{(2\times3)+1}{3} = \frac{7}{3}$

46. $\begin{aligned} 3\frac{7}{20} &= 3\frac{35}{100} \\ +7\frac{3}{25} &= 7\frac{12}{100} \\ \hline 10\frac{47}{100} \end{aligned}$

47. $\begin{aligned} 3\frac{1}{5} &= \frac{16}{5} \\ -2\frac{2}{5} &= \frac{12}{5} \\ \hline \frac{4}{5} \end{aligned}$ inch

## Multiplication with Mixed Numerals (page 16)

1. $9 \times \frac{5}{6} = \frac{9}{1} \times \frac{5}{\cancel{6}} = \frac{15}{2}$ or $7\frac{1}{2}$

2. $5\frac{1}{3} \times 1\frac{4}{5} = \frac{16}{3} \times \frac{\cancel{9}}{5} = \frac{48}{5}$ or $9\frac{3}{5}$

3. $\frac{1}{2} \times 4\frac{2}{5} = \frac{1}{\cancel{2}} \times \frac{\cancel{22}}{5} = \frac{11}{5}$ or $2\frac{1}{5}$

33

4. $\frac{1}{3} \times 1\frac{1}{7} = \frac{1}{3} \times \frac{8}{7} = \frac{8}{21}$

5. $\frac{2}{7} \times 4\frac{1}{6} = \frac{2}{7} \times \frac{25}{6} = \frac{25}{21}$ or $1\frac{4}{21}$

6. $\frac{4}{5} \times 5\frac{1}{4} = \frac{4}{5} \times \frac{21}{4} = \frac{21}{5}$ or $4\frac{1}{5}$

7. $2\frac{3}{5} \times \frac{5}{6} = \frac{13}{5} \times \frac{5}{6} = \frac{13}{6}$ or $2\frac{1}{6}$

8. $1\frac{5}{6} \times \frac{5}{9} = \frac{11}{6} \times \frac{5}{9} = \frac{55}{54}$ or $1\frac{1}{54}$

9. $3\frac{1}{4} \times \frac{2}{5} = \frac{13}{4} \times \frac{2}{5} = \frac{13}{10}$ or $1\frac{3}{10}$

10. $4\frac{3}{4} \times \frac{1}{3} = \frac{19}{4} \times \frac{1}{3} = \frac{19}{12}$ or $1\frac{7}{12}$

11. $7\frac{5}{8} \times \frac{4}{5} = \frac{61}{8} \times \frac{4}{5} = \frac{61}{10}$ or $6\frac{1}{10}$

12. $\frac{5}{8} \times 5 = \frac{5}{8} \times \frac{5}{1} = \frac{25}{8}$ or $3\frac{1}{8}$

13. $\frac{5}{6} \times 8 = \frac{5}{6} \times \frac{8}{1} = \frac{20}{3}$ or $6\frac{2}{3}$

14. $\frac{3}{5} \times 20 = \frac{3}{5} \times \frac{20}{1} = 12$

15. $7 \times \frac{2}{3} = \frac{7}{1} \times \frac{2}{3} = \frac{14}{3}$ or $4\frac{2}{3}$

16. $1\frac{3}{4} \times 7 = \frac{7}{4} \times \frac{7}{1} = \frac{49}{4}$ or $12\frac{1}{4}$

17. $1\frac{1}{5} \times 4 = \frac{6}{5} \times \frac{4}{1} = \frac{24}{5}$ or $4\frac{4}{5}$

18. $8\frac{2}{3} \times 6 = \frac{26}{3} \times \frac{6}{1} = 52$

19. $3\frac{1}{5} \times 6 = \frac{16}{5} \times \frac{6}{1} = \frac{96}{5}$ or $19\frac{1}{5}$

20. $\frac{7}{8} \times 1\frac{3}{7} = \frac{7}{8} \times \frac{10}{7} = \frac{5}{4}$ or $1\frac{1}{4}$

21. $\frac{2}{5} \times 1\frac{1}{4} = \frac{2}{5} \times \frac{5}{4} = \frac{1}{2}$

22. $1\frac{1}{3} \times \frac{9}{10} = \frac{4}{3} \times \frac{9}{10} = \frac{6}{5}$ or $1\frac{1}{5}$

23. $1\frac{3}{5} \times \frac{3}{4} = \frac{8}{5} \times \frac{3}{4} = \frac{6}{5}$ or $1\frac{1}{5}$

24. $4\frac{3}{8} \times 2 = \frac{35}{8} \times \frac{2}{1} = \frac{35}{4}$ or $8\frac{3}{4}$

25. $7 \times 3\frac{1}{2} = \frac{7}{1} \times \frac{7}{2} = \frac{49}{2}$ or $24\frac{1}{2}$

26. $4 \times 1\frac{1}{3} = \frac{4}{1} \times \frac{4}{3} = \frac{16}{3}$ or $5\frac{1}{3}$

27. $2\frac{1}{8} \times 1\frac{1}{3} = \frac{17}{8} \times \frac{4}{3} = \frac{17}{6}$ or $2\frac{5}{6}$

28. $1\frac{1}{4} \times 1\frac{3}{5} = \frac{5}{4} \times \frac{8}{5} = 2$

29. $1\frac{3}{4} \times 2 = \frac{7}{4} \times \frac{2}{1} = \frac{7}{2}$ or $3\frac{1}{2}$

30. $1\frac{5}{7} \times 2\frac{5}{8} = \frac{12}{7} \times \frac{21}{8} = \frac{9}{2}$ or $4\frac{1}{2}$

31. $1\frac{1}{5} \times 3\frac{3}{4} = \frac{6}{5} \times \frac{15}{4} = \frac{9}{2}$ or $4\frac{1}{2}$

32. $4\frac{7}{8} \times 1\frac{1}{6} = \frac{39}{8} \times \frac{7}{6} = \frac{91}{16}$ or $5\frac{11}{16}$

33. $\frac{9}{16} \times \frac{3}{8} = \frac{27}{128}$

## Division of Fractions (pages 17–18)

1. $\frac{15}{2}$ or $7\frac{1}{2}$    2. 12    3. $\frac{1}{2}$    4. $\frac{1}{16}$    5. $\frac{10}{7}; \frac{4}{7}$    6. $\frac{5}{3}; \frac{5}{12}$    7. $6 \div \frac{1}{6} = \frac{6}{1} \cdot \frac{6}{1} = 36$    8. $5 \div \frac{1}{3} = \frac{5}{1} \times \frac{3}{1} = 15$

9. $4 \div \frac{1}{3} = \frac{4}{1} \times \frac{3}{1} = 12$    10. $7 \div \frac{1}{4} = \frac{7}{1} \times \frac{4}{1} = 28$    11. $8 \div \frac{2}{5} = \frac{8}{1} \times \frac{5}{2} = 20$    12. $10 \div \frac{2}{7} = \frac{10}{1} \times \frac{7}{2} = 35$

13. $12 \div \frac{16}{11} = \frac{12}{1} \times \frac{11}{16} = 22$    14. $12 \div \frac{2}{3} = \frac{12}{1} \times \frac{3}{2} = 18$    15. $\frac{1}{2} \div 5 = \frac{1}{2} \times \frac{1}{5} = \frac{1}{10}$    16. $\frac{1}{4} \div 12 = \frac{1}{4} \times \frac{1}{12} = \frac{1}{48}$

17. $\frac{1}{3} \div 6 = \frac{1}{3} \times \frac{1}{6} = \frac{1}{18}$    18. $\frac{1}{7} \div 2 = \frac{1}{7} \times \frac{1}{2} = \frac{1}{14}$    19. $\frac{9}{15} \div \frac{3}{15} = \frac{9}{15} \times \frac{15}{3} = 3$    20. $\frac{7}{8} \div \frac{7}{8} = \frac{7}{8} \times \frac{8}{7} = 1$

21. $\frac{4}{7} \div \frac{9}{28} = \frac{4}{7} \times \frac{28}{9} = \frac{16}{9}$ or $1\frac{7}{9}$    22. $\frac{4}{9} \div \frac{4}{5} = \frac{4}{9} \times \frac{5}{4} = \frac{5}{9}$    23. $\frac{1}{4} \div \frac{5}{8} = \frac{1}{4} \times \frac{8}{5} = \frac{2}{5}$    24. $\frac{1}{7} \div \frac{5}{6} = \frac{1}{7} \times \frac{6}{5} = \frac{6}{35}$

25. $\frac{1}{2} \div \frac{5}{12} = \frac{1}{2} \times \frac{12}{5} = \frac{6}{5}$ or $1\frac{1}{5}$    26. $\frac{1}{16} \div \frac{3}{4} = \frac{1}{16} \times \frac{4}{3} = \frac{1}{12}$    27. $\frac{3}{4} \div \frac{3}{8} = \frac{3}{4} \times \frac{8}{3} = 2$    28. $\frac{3}{5} \div \frac{3}{4} = \frac{3}{5} \times \frac{4}{3} = \frac{4}{5}$

29. $\frac{7}{8} \div \frac{3}{8} = \frac{7}{8} \times \frac{8}{3} = \frac{7}{3}$ or $2\frac{1}{3}$    30. $8 \div \frac{1}{3} = \frac{8}{1} \times \frac{3}{1} = 24$    31. $6 \div \frac{4}{5} = \frac{6}{1} \times \frac{5}{4} = \frac{15}{2}$ or $7\frac{1}{2}$

32. $9 \div \frac{5}{8} = \frac{9}{1} \times \frac{8}{5} = \frac{72}{5}$ or $14\frac{2}{5}$    33. $5 \div \frac{2}{5} = \frac{5}{1} \times \frac{5}{2} = \frac{25}{2}$ or $12\frac{1}{2}$    34. $\frac{5}{12} \div \frac{2}{3} = \frac{5}{12} \times \frac{3}{1} = \frac{5}{8}$

35. $7 \div \frac{1}{2} = \frac{7}{1} \times \frac{2}{1} = 14$   36. $6 \div \frac{3}{4} = \frac{6}{1} \times \frac{4}{3} = 8$   37. $\frac{9}{10} \div 3 = \frac{9}{10} \times \frac{1}{3} = \frac{3}{10}$   38. $\frac{2}{3} \div \frac{3}{4} = \frac{2}{3} \times \frac{4}{3} = \frac{8}{9}$

39. $5 \div \frac{1}{8} = \frac{5}{1} \times \frac{8}{1} = 40$   40. $2 \div \frac{5}{6} = \frac{2}{1} \times \frac{6}{5} = \frac{12}{5}$ or $2\frac{2}{5}$   41. $\frac{7}{10} \div \frac{14}{25} = \frac{7}{10} \times \frac{25}{14} = \frac{5}{4}$ or $1\frac{1}{4}$

42. $\frac{4}{7} \div \frac{1}{3} = \frac{4}{7} \times \frac{3}{1} = \frac{12}{7}$ or $1\frac{5}{7}$   43. $\frac{2}{7} \div 5 = \frac{2}{7} \times \frac{1}{5} = \frac{2}{35}$   44. $\frac{5}{6} \div 2 = \frac{5}{6} \times \frac{1}{2} = \frac{5}{12}$   45. $\frac{5}{8} \div \frac{5}{6} = \frac{5}{8} \times \frac{6}{5} = \frac{3}{4}$

46. $\frac{4}{5} \div 4 = \frac{4}{5} \times \frac{1}{4} = \frac{1}{5}$   47. $\frac{1}{6} \div \frac{2}{3} = \frac{1}{6} \times \frac{3}{2} = \frac{1}{4}$   48. $\frac{7}{10} \div \frac{2}{3} = \frac{7}{10} \times \frac{3}{2} = \frac{21}{20}$ or $1\frac{1}{20}$

49. $\frac{7}{9} \div \frac{7}{12} = \frac{7}{9} \times \frac{12}{7} = \frac{4}{3}$ or $1\frac{1}{3}$   50. $\frac{3}{4} \div \frac{6}{7} = \frac{3}{4} \times \frac{7}{6} = \frac{7}{8}$   51. $\frac{8}{9} \div \frac{4}{9} = \frac{8}{9} \times \frac{9}{4} = 2$

52. $\frac{5}{8} \div \frac{5}{6} = \frac{5}{8} \times \frac{6}{5} = \frac{3}{4}$   53. $\frac{7}{20} \div \frac{28}{35} = \frac{7}{20} \times \frac{35}{28} = \frac{7}{16}$   54. $\frac{11}{16} \div \frac{3}{8} = \frac{11}{16} \times \frac{8}{3} = \frac{11}{6}$ or $1\frac{5}{6}$

55. $\frac{1}{12} \div \frac{2}{9} = \frac{1}{12} \times \frac{9}{2} = \frac{3}{8}$ inch long dwarf bee   56. $\frac{4}{5} \div \frac{1}{10} = \frac{4}{5} \times \frac{10}{1} = 8$ honeybees

57. $\frac{3}{10} \div \frac{3}{5} = \frac{3}{10} \times \frac{5}{3} = \frac{1}{2}$ of the fertilizer is nitrogen   58. $\frac{2}{3} \div 4 = \frac{2}{3} \times \frac{1}{4} = \frac{2}{12}$ or $\frac{1}{6}$ of the fertilizer is phosphorus

59. $\frac{3}{5} \div \frac{1}{10} = \frac{3}{5} \times \frac{10}{1} = 6$ bees   60. $\frac{5}{6} \div \frac{1}{12} = \frac{5}{6} \times \frac{12}{1} = 10$ strips

## Division with Mixed Numerals (pages 19–20)

1. $\frac{2}{3}$   2. $\frac{7}{20}$   3. $\frac{3}{7}$; $\frac{12}{7}$ or $1\frac{5}{7}$   4. $\frac{1}{6}$; $\frac{3}{5}$

5. $\frac{5}{8}$; $\frac{3}{4}$   6. $\frac{8}{15}$; $\frac{6}{5}$ or $1\frac{1}{5}$   7. $5 \div 3\frac{1}{3} = \frac{5}{1} \div \frac{10}{3} = \frac{5}{1} \times \frac{3}{10} = \frac{3}{2}$ or $1\frac{1}{2}$

8. $6\frac{1}{4} \div 5 = \frac{25}{4} \div \frac{5}{1} = \frac{25}{4} \times \frac{1}{5} = \frac{5}{4}$ or $1\frac{1}{4}$   9. $4 \div 1\frac{1}{3} = \frac{4}{1} \div \frac{4}{3} = \frac{4}{1} \times \frac{3}{4} = 3$

10. $1\frac{5}{8} \div 2 = \frac{13}{8} \div \frac{2}{1} = \frac{13}{8} \times \frac{1}{2} = \frac{13}{16}$   11. $1\frac{3}{5} \div \frac{5}{8} = \frac{8}{5} \div \frac{5}{8} = \frac{8}{5} \times \frac{8}{5} = \frac{64}{25} = 2\frac{14}{25}$

12. $3\frac{1}{8} \div 1\frac{1}{4} = \frac{25}{8} \div \frac{5}{4} = \frac{25}{8} \times \frac{4}{5} = \frac{5}{2}$ or $2\frac{1}{2}$   13. $3 \div 4\frac{1}{2} = \frac{3}{1} \div \frac{9}{2} = \frac{3}{1} \times \frac{2}{9} = \frac{2}{3}$

14. $2\frac{3}{4} \div \frac{5}{6} = \frac{11}{4} \div \frac{5}{6} = \frac{11}{4} \times \frac{6}{5} = \frac{33}{10}$ or $3\frac{3}{10}$   15. $\frac{5}{6} \div 1\frac{1}{9} = \frac{5}{6} \div \frac{10}{9} = \frac{5}{6} \times \frac{9}{10} = \frac{3}{4}$

16. $2\frac{2}{5} \div \frac{3}{10} = \frac{12}{5} \div \frac{3}{10} = \frac{12}{5} \times \frac{10}{3} = 8$   17. $\frac{8}{9} \div 2\frac{2}{5} = \frac{8}{9} \div \frac{12}{5} = \frac{8}{9} \times \frac{5}{12} = \frac{10}{27}$

18. $1\frac{2}{5} \div 2\frac{2}{3} = \frac{7}{5} \div \frac{8}{3} = \frac{7}{5} \times \frac{3}{8} = \frac{21}{40}$   19. $\frac{9}{10} \div 5\frac{2}{5} = \frac{9}{10} \div \frac{27}{5} = \frac{9}{10} \times \frac{5}{27} = \frac{1}{6}$

20. $1\frac{1}{4} \div 4\frac{1}{2} = \frac{5}{4} \div \frac{9}{2} = \frac{5}{4} \times \frac{2}{9} = \frac{5}{18}$   21. $4\frac{2}{3} \div 1\frac{3}{5} = \frac{14}{3} \div \frac{8}{5} = \frac{14}{3} \times \frac{5}{8} = \frac{35}{12}$ or $2\frac{11}{12}$

**22.** $3\frac{1}{2} \div 5\frac{1}{2} = \frac{7}{2} \div \frac{11}{2} = \frac{7}{2} \times \frac{2}{11} = \frac{7}{11}$

**23.** $3\frac{1}{5} \div 1\frac{1}{3} = \frac{16}{5} \div \frac{4}{3} = \frac{16}{5} \times \frac{3}{4} = \frac{12}{5}$ or $2\frac{2}{5}$

**24.** $1\frac{1}{4} \div 7\frac{1}{2} = \frac{5}{4} \div \frac{15}{2} = \frac{5}{4} \times \frac{2}{15} = \frac{1}{6}$

**25.** $4\frac{3}{4} \div 1\frac{7}{8} = \frac{19}{4} \div \frac{15}{8} = \frac{19}{4} \times \frac{8}{15} = \frac{38}{15}$ or $2\frac{8}{15}$

**26.** $5\frac{1}{2} \div 2\frac{3}{4} = \frac{11}{2} \div \frac{11}{4} = \frac{11}{2} \times \frac{4}{11} = 2$

**27.** $9\frac{3}{5} \div 4\frac{1}{5} = \frac{48}{5} \div \frac{21}{5} = \frac{48}{5} \times \frac{5}{21} = \frac{16}{7}$ or $2\frac{2}{7}$

**28.** $1\frac{2}{3} \div 6\frac{3}{7} = \frac{5}{3} \div \frac{45}{7} = \frac{5}{3} \times \frac{7}{45} = \frac{7}{27}$

**29.** $2\frac{3}{5} \div 11\frac{4}{5} = \frac{13}{5} \div \frac{59}{5} = \frac{13}{5} \times \frac{5}{59} = \frac{13}{59}$

**30.** $2\frac{1}{9} \div 3\frac{1}{3} = \frac{19}{9} \div \frac{10}{3} = \frac{19}{9} \times \frac{3}{10} = \frac{19}{30}$

**31.** $2\frac{1}{4} \div 2\frac{1}{4} = \frac{9}{4} \div \frac{9}{4} = \frac{9}{4} \times \frac{4}{9} = 1$

**32.** $1\frac{1}{8} \div 3\frac{3}{8} = \frac{9}{8} \div \frac{27}{8} = \frac{9}{8} \times \frac{8}{27} = \frac{1}{3}$

**33.** $4\frac{5}{9} \div 1\frac{2}{9} = \frac{41}{9} \div \frac{11}{9} = \frac{41}{9} \times \frac{9}{11} = \frac{41}{11}$ or $3\frac{8}{11}$

**34.** $1\frac{1}{9} \div 2\frac{4}{9} = \frac{10}{9} \div \frac{22}{9} = \frac{10}{9} \times \frac{9}{22} = \frac{5}{11}$

**35.** $1\frac{1}{5} \div 2\frac{3}{5} = \frac{6}{5} \div \frac{13}{5} = \frac{6}{5} \times \frac{5}{13} = \frac{6}{13}$

**36.** $6\frac{3}{5} \div 2\frac{1}{5} = \frac{33}{5} \div \frac{11}{5} = \frac{33}{5} \times \frac{5}{11} = 3$

**37.** $3\frac{9}{10} \div 2\frac{13}{15} = \frac{39}{10} \div \frac{43}{15} = \frac{39}{10} \times \frac{15}{43} = \frac{117}{86}$ or $1\frac{31}{86}$

**38.** $10\frac{1}{7} \div 3\frac{4}{7} = \frac{71}{7} \div \frac{25}{7} = \frac{71}{7} \times \frac{7}{25} = \frac{71}{25}$ or $2\frac{21}{25}$

**39.** $2\frac{2}{7} \div 2\frac{1}{2} = \frac{16}{7} \div \frac{5}{2} = \frac{16}{7} \times \frac{2}{5} = \frac{32}{35}$

**40.** $5\frac{1}{21} \div 2\frac{21}{35} = \frac{106}{21} \div \frac{91}{35} = \frac{106}{21} \times \frac{35}{91} = \frac{530}{273}$ or $1\frac{257}{273}$

**41.** $3\frac{2}{7} \div 2\frac{9}{21} = \frac{23}{7} \div \frac{51}{21} = \frac{23}{7} \times \frac{21}{51} = \frac{69}{51}$ or $1\frac{18}{51}$

**42.** $3\frac{3}{5} \div 2\frac{10}{15} = \frac{18}{5} \div \frac{40}{15} = \frac{18}{5} \times \frac{15}{40} = \frac{27}{20}$ or $1\frac{7}{20}$

**43.** $5\frac{1}{3} \div 1\frac{1}{9} = \frac{16}{3} \div \frac{10}{9} = \frac{16}{3} \times \frac{9}{10} = \frac{24}{5}$ or $4\frac{4}{5}$

**44.** $1\frac{3}{8} \div 3\frac{1}{16} = \frac{11}{8} \div \frac{49}{16} = \frac{11}{8} \times \frac{16}{49} = \frac{22}{49}$

**45.** $3\frac{5}{8} \div 7\frac{11}{16} = \frac{29}{8} \div \frac{123}{16} = \frac{29}{8} \times \frac{16}{123} = \frac{58}{123}$

**46.** $7\frac{1}{4} \div 3\frac{7}{8} = \frac{29}{4} \div \frac{31}{8} = \frac{29}{4} \times \frac{8}{31} = \frac{58}{31}$ or $1\frac{27}{31}$

**47.** $2\frac{2}{9} \div 1\frac{7}{18} = \frac{20}{9} \div \frac{25}{18} = \frac{20}{9} \times \frac{18}{25} = \frac{8}{5}$ or $1\frac{3}{5}$

**48.** $2\frac{2}{3} \div 1\frac{1}{15} = \frac{8}{3} \div \frac{16}{15} = \frac{8}{3} \times \frac{15}{16} = \frac{5}{2}$ or $2\frac{1}{2}$

**49.** $10 \div 2\frac{1}{5} = \frac{10}{1} \div \frac{11}{5} = \frac{10}{1} \times \frac{5}{11} = \frac{50}{11}$ or $4\frac{6}{11}$

**50.** $2\frac{2}{3} \div 4 = \frac{8}{3} \div \frac{4}{1} = \frac{8}{3} \times \frac{1}{4} = \frac{2}{3}$

**51.** $3 \div 1\frac{1}{8} = \frac{3}{1} \div \frac{9}{8} = \frac{3}{1} \times \frac{8}{9} = \frac{8}{3}$ or $2\frac{2}{3}$

**52.** $\frac{7}{8} \div 1\frac{1}{4} = \frac{7}{8} \div \frac{5}{4} = \frac{7}{8} \times \frac{4}{5} = \frac{7}{10}$

**53.** $1\frac{1}{6} \div \frac{2}{3} = \frac{7}{6} \div \frac{2}{3} = \frac{7}{6} \times \frac{3}{2} = \frac{7}{4}$

**54.** $\frac{9}{10} \div 2\frac{4}{5} = \frac{9}{10} \div \frac{14}{5} = \frac{9}{10} \times \frac{5}{14} = \frac{9}{28}$

**55.** $3\frac{1}{5} \div 1\frac{1}{3} = \frac{16}{5} \div \frac{4}{3} = \frac{16}{5} \times \frac{3}{4} = \frac{12}{5}$ or $2\frac{2}{5}$

**56.** $1\frac{2}{9} \div 3\frac{1}{5} = \frac{11}{9} \div \frac{16}{5} = \frac{11}{9} \times \frac{5}{16} = \frac{55}{144}$

**57.** $6\frac{1}{2} \div 4\frac{1}{4} = \frac{13}{2} \div \frac{17}{4} = \frac{13}{2} \times \frac{\overset{2}{4}}{17} = \frac{26}{17}$ or $1\frac{9}{17}$

**58.** $1 \div 3\frac{1}{3} = \frac{1}{1} \div \frac{10}{3} = \frac{1}{1} \times \frac{3}{10} = \frac{3}{10}$ lb per foot per row of spinach

**59.** $5\frac{1}{3} \div 6\frac{2}{3} = \frac{16}{3} \div \frac{20}{3} = \frac{\overset{4}{16}}{3} \times \frac{\overset{1}{3}}{\underset{5}{20}} = \frac{4}{5}$ lb per foot per row of swiss chard

**60.** $3\frac{1}{2} \div 5 = \frac{7}{2} \div \frac{5}{1} = \frac{7}{2} \times \frac{1}{5} = \frac{7}{10}$ lb of oranges per child

**61.** $2\frac{7}{9} \div 6\frac{2}{3} = \frac{25}{9} \div \frac{20}{3} = \frac{\overset{5}{25}}{\underset{3}{9}} \times \frac{\overset{1}{3}}{\underset{4}{20}} = \frac{5}{12}$ mile per minute (dolphin)

**62.** $2\frac{2}{9} \div 3\frac{1}{3} = \frac{20}{9} \div \frac{10}{3} = \frac{\overset{2}{20}}{\underset{3}{9}} \times \frac{\overset{1}{3}}{\underset{1}{10}} = \frac{2}{3}$ mile per minute (owl)

**63.** $6\frac{1}{2} \div 5\frac{3}{4} = \frac{13}{2} \div \frac{23}{4} = \frac{13}{2} \times \frac{\overset{2}{4}}{23} = \frac{26}{23}$ or $1\frac{3}{23}$ lb per foot per row of peas

**64.** $3 \div 3\frac{3}{5} = \frac{3}{1} \div \frac{18}{5} = \frac{\overset{1}{3}}{1} \times \frac{5}{\underset{6}{18}} = \frac{5}{6}$ mile per minute (dragonfly)

**65.** $1\frac{5}{6} \div 5\frac{1}{2} = \frac{11}{6} \div \frac{11}{2} = \frac{\overset{1}{11}}{\underset{3}{6}} \times \frac{\overset{1}{2}}{\underset{1}{11}} = \frac{1}{3}$ mile per minute (whale)

## Circumference of a Circle (page 21)

**1.** $C = \frac{2}{1} \times \frac{22}{7} \times \frac{\overset{2}{14}}{1}$  Multiply by 2 since the radius is given.
$C = 88$ cm

**2.** $C = \frac{22}{7} \times \frac{\overset{6}{42}}{1}$
$C = 132$ m

**3.** $C = \frac{2}{1} \times \frac{22}{7} \times \frac{\overset{4}{28}}{1}$  Multiply by 2 since the radius is given.
$C = 176$ ft

**4.** $C = \frac{22}{7} \times \frac{\overset{3}{21}}{1}$
$C = 66$ m

**5.** $C = \frac{22}{7} \times \frac{\overset{9}{63}}{1}$
$C = 198$ ft

**6.** $C = \frac{22}{7} \times \frac{\overset{11}{77}}{1}$
$C = 242$ m

**7.** $C = \frac{22}{7} \times \frac{\overset{18}{126}}{1}$
$C = 396$ mm

**8.** $C = \frac{22}{7} \times \frac{\overset{212}{1484}}{1}$
$C = 4664$ ft

**9.** $C = \frac{22}{7} \times \frac{\overset{33}{231}}{1}$
$C = 726$ cm

**10.** $C = \frac{22}{7} \times \frac{\overset{64}{448}}{1}$
$C = 1408$ in.

**11.** $C = \frac{22}{7} \times \frac{\overset{41}{287}}{1}$
$C = 902$ cm

**12.** $C = \frac{22}{7} \times \frac{\overset{26}{182}}{1}$
$C = 572$ in.

**13.** $C = \frac{22}{7} \times \frac{47}{1}$
$C = \frac{1034}{7}$ or $147\frac{5}{7}$ in.

**14.** $C = \frac{22}{7} \times \frac{9}{1}$
$C = \frac{198}{7}$ or $28\frac{2}{7}$ ft

**15.** $C = \frac{22}{7} \times \frac{5}{1}$
$C = \frac{110}{7}$ or $15\frac{5}{7}$ in.

**16.** $C = \frac{22}{7} \times \frac{12}{1}$
$C = \frac{264}{7}$ or $37\frac{5}{7}$ cm

## Addition of Integers (page 22)

**1.** –7  **2.** +1 or 1  **3.** +6 or 6  **4.** +3 or 3  **5.** –2  **6.** +8 or 8  **7.** –5  **8.** 0

**9.** ←——————  **10.** ——————→ (with ←—— below)  **11.** ←—— (with ——→ below)

**12.** $-7 + -3 = -10$  **13.** $-7 + 5 = -2$ since –7 has the longest arrow by 2.  **14.** $9 + 12 = 21$

**15.** $8 + -8 = 0$  **16.** $0 + -9 = -9$  **17.** $-5 + 9 = 4$ since 9 has the longest arrow by 4.  **18.** $-4 + -11 = -15$

**19.** $9 + -7 = 2$ since 9 has the longest arrow by 2.  **20.** $-10 + 8 = -2$ since –10 has the longest arrow by 2.

**21.** $4 + -4 = 0$  **22.** $-7 + 0 = -7$  **23.** $-4 + -6 = -10$

## Subtraction of Integers (page 23)

**1.** –7  **2.** 7  **3.** 2  **4.** –8  **5.** 11  **6.** 5  **7.** –14  **8.** 12  **9.** 17  **10.** –1

**11.** $7 + -5 = 2$; add the opposite of 5 or –5.  **12.** $-9 + -6 = -15$; add the opposite of 6 or –6.

**13.** $6 + 4 = 10$; add the opposite of –4 or 4.  **14.** $-8 + -3 = -11$; add the opposite of 3 or –3.

**15.** $3 + 2 = 5$; add the opposite of –2 or 2.  **16.** $0 + 9 = 9$; add the opposite of –9 or 9.

**17.** 2  **18.** 2  **19.** –2, since $9 + -11 = -2$.  **20.** 0, since $-8 + 8 = 0$.  **21.** 1  **22.** –14  **23.** 9  **24.** –10

**25.** 5, since 4 + 1 = 5.    **26.** −17, since −11 + −6 = −17.    **27.** 8, −3 + 11 = 8    **28.** 6    **29.** 1.9
**30.** 15, since 7 + 8 = 15.    **31.** −9.1, since −5.4 + −3.7 = −9.1    **32.** −8, since −3 + −5 = −8.

## Multiplying Integers (page 24)

**1.** 16, same signs
**2.** 8, same signs
**3.** 27, same signs
**4.** 48, same signs
**5.** −8, different signs
**6.** −33, different signs
**7.** −35, different signs
**8.** −63, different signs
**9.** 28, same signs
**10.** −25, different signs
**11.** −49, different signs
**12.** 15, same signs
**13.** −12, different signs
**14.** 36, same signs
**15.** 6, same signs
**16.** −24, different signs
**17.** −140, different signs
**18.** 336, same signs
**19.** −136, different signs
**20.** 165, same signs
**21.** 625, same signs
**22.** −960, different signs
**23.** 300, same signs
**24.** −1750, different signs

## Dividing Integers (page 25)

**1.** 16, same signs
**2.** 20, same signs
**3.** 3, same signs
**4.** 7, same signs
**5.** −7, different signs
**6.** −6, different signs
**7.** −27, different signs
**8.** −4, different signs
**9.** 3, same signs
**10.** −9, different signs
**11.** 9, same signs
**12.** −10, different signs
**13.** −5, different signs
**14.** 8, same signs
**15.** −32, different signs
**16.** 28, same signs
**17.** 8, same signs
**18.** −6, different signs
**19.** 12, same signs
**20.** −9, different signs
**21.** −24, different signs
**22.** 12, same signs
**23.** −15, different signs
**24.** 56, same signs

# DECIMALS

## Decimal Place Value

The chart shows the place value of each digit in 8653.427. The chart can be extended in either direction.

Note that the 2 is in the hundredths position. The digit 2 and its place value name the number 0.02.

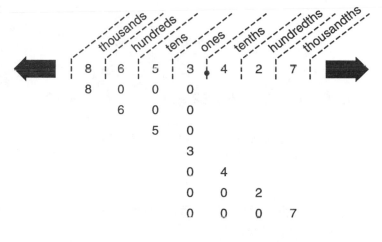

$$\underset{\text{digit}}{\underline{2}} \times \underset{\text{place value}}{\underline{0.01}} = \underset{\text{number}}{\underline{0.02}} \text{ or } \frac{2}{100}$$

*Name the place value for each of the following digits in 4527.386.*

| | | | | | |
|---|---|---|---|---|---|
| **1.** the 7 | **2.** the 4 | **3.** the 6 | **4.** the 5 | **5.** the 3 | **6.** the 2 |

*Name the place value for each of the following digits in 6758.9142*

| | | | | | |
|---|---|---|---|---|---|
| **7.** the 9 | **8.** the 1 | **9.** the 7 | **10.** the 2 | **11.** the 5 | **12.** the 4 |
| **13.** the 6 | **14.** the 8 | | | | |

*What number is named by each of the following digits in 4527.386?*

| | | | | | |
|---|---|---|---|---|---|
| **15.** the 7 | **16.** the 4 | **17.** the 6 | **18.** the 5 | **19.** the 3 | **20.** the 2 |

*What number is named by each of the following digits in 7238.05461?*

| | | | | | |
|---|---|---|---|---|---|
| **21.** the 7 | **22.** the 5 | **23.** the 6 | **24.** the 8 | **25.** the 3 | **26.** the 4 |
| **27.** the 1 | **28.** the 2 | | | | |

*What number is named by each of the following digits in 4269.1853?*

| | | | | | |
|---|---|---|---|---|---|
| **29.** the 2 | **30.** the 9 | **31.** the 1 | **32.** the 5 | **33.** the 6 | **34.** the 3 |
| **35.** the 4 | **36.** the 8 | | | | |

*Name the place value of the underlined digit.*

| | | | | | |
|---|---|---|---|---|---|
| **37.** 87.<u>3</u> | **38.** 9.56<u>7</u> | **39.** 0.61<u>8</u> | **40.** 24.<u>7</u>5 | **41.** 2.00<u>1</u> | **42.** 22.4<u>16</u> |
| **43.** <u>6</u>71.2 | **44.** 14.<u>58</u> | **45.** 5<u>2</u>3.9 | **46.** 0.08<u>8</u> | **47.** 1.7<u>6</u> | **48.** 3.9<u>51</u> |
| **49.** <u>3</u>.1 | **50.** 0.39<u>5</u> | **51.** 85.4<u>6</u> | **52.** 2<u>4</u>.53 | **53.** 3<u>32</u>.5 | **54.** 660.<u>2</u> |
| **55.** 41.<u>8</u> | **56.** 2.<u>4</u> | **57.** 7.<u>2</u> | **58.** 5.04<u>7</u> | **59.** 1.2<u>6</u> | **60.** <u>1</u>92.9 |
| **61.** <u>5</u>76.8 | **62.** 0.1<u>3</u> | **63.** 23.1<u>7</u> | | | |

Decimal numbers can be written in two ways, the standard form and the word form. Using **27.51**, the standard form is given. To write in words, it is written *twenty-seven and fifty-one hundredths*.

*Copy each of the following in words.*

**64.** 24.71      **65.** 17.94      **66.** 33.25      **67.** 11.394      **68.** 37.98

**69.** 0.704      **70.** 82.951      **71.** 14.306      **72.** 48.8192      **73.** 13.325

Numbers are written in both standard form and in word form when writing checks.

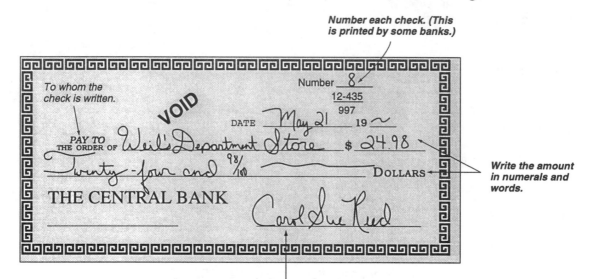

Number each check. (This is printed by some banks.)

To whom the check is written.

Write the amount in numerals and words.

Signature of the check writer. Do not sign until you are sure the check is written correctly.

*Copy and write each check amount in words.*

**74.** $7.65 _____ Dollars

**75.** $15.00 _____ Dollars

**76.** $63.59 _____ Dollars

**77.** $45.07 _____ Dollars

**78.** $21.08 _____ Dollars

**79.** $125.14 _____ Dollars

**80.** $2.98 _____ Dollars

**81.** $376.23 _____ Dollars

**82.** $50.00 _____ Dollars

**83.** $2948.31 _____ Dollars

**84.** $84.72 _____ Dollars

# Comparing Decimals

The shading in the figures shows that 0.1 and 0.10 are equivalent. That is, 0.1 = 0.10.

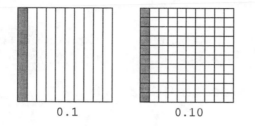

0.1          0.10

> **You can annex zeros to the right of a decimal without a change in value.**

Compare 1.3 and 1.35. First, annex a zero to 1.3. Now, 1.30 and 1.35 have the same number of decimal places. Compare as with whole numbers.

**130 < 135 so 1.30 < 1.35. Therefore, 1.3 < 1.35.**

*True or False.*

| | | |
|---|---|---|
| **1.** 0.40 = 0.400 | **2.** 0.30 = 0.03 | **3.** 1.0 = 0.1 |
| **4.** 20.02 − 20.20 | **5.** 0.18 = 0.180 | **6.** 7.1 = 7.10 |
| **7.** 33.3 = 33.33 | **8.** 14 = 14.00 | **9.** 0.47 = 0.470 |
| **10.** 0.08 < 0.8 | **11.** 0.07 > 0.5 | **12.** 0.05 = 0.55 |
| **13.** 0.58 = 5.8 | **14.** 0.9 > 0.7 | **15.** 6 < 6.001 |
| **16.** 10.002 > 10.02 | **17.** 8.05 < 8.002 | **18.** 0.003 > 0.030 |
| **19.** 0.5 = 0.52 | **20.** 0.22 < 0.32 | **21.** 0.55 > 0.13 |
| **22.** 4.14 < 5.15 | **23.** 2.7 > 7.2 | **24.** 0.3 = 0.300 |
| **25.** 0.71 > 0.7 | **26.** 0.44 < 0.99 | **27.** 0.36 < 0.65 |
| **28.** 8.1 < 6.83 | **29.** 0.09 < 0.19 | **30.** 5.4 < 5.44 |

**31.** The winning times in the women's Olympic 200-meter dash are: 24, 22, 24.6, 23.7, 22.4, 22.37, 22.7, and 23.4. List the times in order from least to greatest.

*Use <, >, or = to make true sentences.*

| | | |
|---|---|---|
| **32.** 0.9 _____ 0.90 | **33.** 0.6 _____ 0.7 | **34.** 0.904 _____ 0.909 |
| **35.** 0.09 _____ 0.90 | **36.** 0.06 _____ 0.60 | **37.** 2.03 _____ 3.02 |
| **38.** 0.49 _____ 0.44 | **39.** 0.84 _____ 0.840 | **40.** 6.216 _____ 6.215 |

41

# Rounding Decimals

To round 64.3$\underline{2}$6 to the underlined place-value position, decide if 64.326 is closer to 64.32 or 64.33.

| **Round up if the digit to the right is 5, 6, 7, 8, or 9.** | **Round down if the digit to the right is 0, 1, 2, 3, or 4.** |
|---|---|

To the underlined place-value position, 64.326 is 64.33.

*Round each to the underlined place-value position.*

1. 33.$\underline{6}$9
2. 77.$\underline{3}$4
3. 5.$\underline{4}$8
4. 49.$\underline{6}$8
5. 28.$\underline{4}$7

6. 6.$\underline{6}$39
7. 21.$\underline{4}$4
8. 13.$\underline{5}$9
9. 4.$\underline{9}$37
10. 10.$\underline{2}$7

*Round each to the underlined place-value position.*

11. $\underline{8}$.7659
12. 8.$\underline{7}$648
13. 8.7$\underline{6}$49
14. 8.56$\underline{4}$9
15. 0.$\underline{6}$1

16. 1.7$\underline{8}$9
17. 5.8$\underline{1}$2
18. 39.1$\underline{2}$5
19. 1.8$\underline{5}$5
20. 0.8$\underline{7}$4

21. 0.9$\underline{7}$7
22. 8.2$\underline{1}$6
23. 0.5$\underline{9}$2
24. 21.2$\underline{8}$4
25. 3.9$\underline{7}$3

26. 0.0$\underline{4}$9
27. 6.$\underline{2}$5
28. 2$\underline{3}$.28
29. 6.$\underline{8}$8
30. 0.00$\underline{8}$5

31. 14.$\underline{0}$8
32. 6.01$\underline{0}$9
33. \$6.$\underline{0}$5
34. \$9.2$\underline{6}$7
35. 0.3$\underline{0}$8

*Round to the nearest whole number (or nearest dollar).*

36. 6.4
37. 7.8
38. 0.85
39. 7.78
40. 2.8

41. \$7.95
42. \$2.05
43. 9.4
44. 9.7
45. 7.42

*Round to the nearest tenth.*

46. 32.07
47. 0.28
48. 0.85
49. 0.09
50. 0.16

51. 6.26
52. 0.22
53. 44.98
54. 0.95
55. 0.816

*Round to the nearest hundredth (or nearest cent).*

56. 7.652
57. 3.4363
58. \$8.0455
59. 66.065
60. 0.155

61. 9.069
62. 0.057
63. \$8.251
64. 0.052
65. \$3.095

*Round to the nearest thousandth (or nearest tenth of a cent).*

66. 0.00657
67. 0.0028
68. 0.0887
69. 5.1066
70. 49.0141

71. 6.9704
72. \$0.6095
73. \$0.3254
74. 7.6609
75. \$0.0037

# Addition of Decimals

Always add in each place-value position from least to greatest.

Add $10, $23.45, and $4.52

| Align the decimal point. Then add as with whole numbers. | First, add the **hundredths.** | Now, add the **tenths.** |
|---|---|---|
| $10.00<br>23.45<br>+ 4.52 | 5 + 2 = 7<br>Note: $10 = $10.00<br>$10.00<br>23.45<br>+ 4.52<br>7 | 4 + 5 = 9<br>$10.00<br>23.45<br>+ 4.52<br>97 |

*Add.*

| 1. | 0.7<br>+0.2 | 2. | 3.1<br>+4.2 | 3. | 9<br>+0.4 | 4. | 4.1<br>+4 | 5. | 8.2<br>+0.3 |
|---|---|---|---|---|---|---|---|---|---|
| 6. | 8.4<br>+3.4 | 7. | 3.4<br>+5.5 | 8. | 0.5<br>+4.3 | 9. | 8.6<br>+0.2 | 10. | 0.44<br>+0.12 |
| 11. | 65.1<br>+11.6 | 12. | 3.202<br>+4.681 | 13. | $52.95<br>+ 7.02 | 14. | 4.81<br>+3.03 | 15. | 6.513<br>+2.382 |
| 16. | 3.2<br>1.139<br>+6.04 | 17. | 82.3<br>1.078<br>+4.51 | 18. | 7.5<br>20<br>+2.485 | 19. | 6.23<br>40.5<br>+1.154 | 20. | 71<br>2.51<br>+10.079 |

The following example shows addition of decimals which require renaming.

Add 27.65 and 5.78.

| Align the decimal points and add as with whole numbers. | First, add the **hundredths.** | Next, add the **tenths.** |
|---|---|---|
| 27.65<br>+ 5.78 | 5 + 8 = 13<br>1<br>27.65<br>+ 5.78<br>3<br><br>13 hundredths =<br>1 tenth and 3 hundredths | 1 + 6 + 7 = 14<br>1 1<br>27.65<br>+ 5.78<br>.43<br><br>14 tenths =<br>1 one and 4 tenths |

*Add.*

| 21. | 4.8<br>+7.4 | 22. | $0.26<br>+0.18 | 23. | 3.6<br>+0.8 | 24. | 0.8<br>+2.8 | 25. | 2.75<br>+2.8 |
|---|---|---|---|---|---|---|---|---|---|
| 26. | 0.82<br>+0.92 | 27. | 4.9<br>+3.47 | 28. | 72.8<br>+5.4 | 29. | 45.67<br>+1.4 | 30. | 2.27<br>+3.71 |

**31.**  5.67
    +0.48

**32.**  $0.39
    +7.96

**33.**  9.5
    +9.9

**34.**  8.26
    +7.94

**35.**  4.309
    +3.745

**36.**  1.834
    +0.005

**37.**  948.6
    + 23.5

**38.**  4.16
    +5.33

**39.**  9.75
    +4.38

**40.**  6.17
    +8.07

**41.**  57.4
    +50.9

**42.**  $5.99
    +7.48

**43.**  0.772
    +0.248

**44.**  85.2
    +56.8

**45.**  5.604
    +1.758

**46.**  47.6
    9.287
    +0.427

**47.**  8.97
    54.2
    +6.066

**48.**  4.1
    37.49
    +108.95

**49.**  47
    8.93
    +14.608

**50.**  62.89
    2.7
    +0.623

**51.**  38.06
    12.6
    1.41
    +0.126

**52.**  70.418
    8.1
    +1.011

**53.**  7.016
    87.368
    +3.07

**54.**  55.76
    55.35
    +2.07

**55.**  83.7
    11.77
    +18.4

Add 18.16 + 40.84

First, align decimal points vertically.

Add as with whole numbers in each place-value position from least to greatest.

18.16
+40.84

1 1
18.16
+40.84
59.00

**56.** 55.18 + 16.93

**57.** 2.15 + 0.22

**58.** 0.5 + 3.9 + 8.3

**59.** 12.75 + 0.25

**60.** 1.9 + 2.01

**61.** 48.72 + 6.195 + 0.06

**62.** 98.08 + 16.93

**63.** 1.09 + 3.8 + 17.32 + 0.758

**64.** 2.854 + 7.6 + 3.42

**65.** 23.95 + 16.09 + 14 + 1.82

*Solve.*

**66.** Juan's deductions are $15.70 for F.I.C.A. tax, $52.89 for withholding tax, $3.42 for local tax, and $8.65 for personal deductions. What are his total deductions?

**67.** A very old coin now measures 2.46 cm in diameter. 0.09 cm has been worn away from handling the coin. What was the original size of the coin?

44

# Subtraction of Decimals

Always subtract in each place-value position from least to greatest.

Subtract 42.76 from 126.3.

Align the decimal point and subtract as with whole numbers.

126.30
− 42.76
─────

First, subtract the **hundredths.**

10 − 6 = 4

$$\begin{array}{r} 126.3\overset{2\ 10}{\cancel{0}} \\ -\ 42.76 \\ \hline 4 \end{array}$$

Note: 126.3 = 126.30

*Rename 1 tenth as 10 hundredths.*

Now, subtract the **tenths.**

12 − 7 = 5

$$\begin{array}{r} 12\overset{5\ \ 1210}{6.3\cancel{0}} \\ -\ 42.76 \\ \hline 54 \end{array}$$

*Rename 1 one as 10 tenths.*
*10 tenths + 2 tenths*
*= 12 tenths*

Check subtraction by adding. 83.54 + 42.76 = 126.30

*Subtract.*

| | | | | | | | | | |
|---|---|---|---|---|---|---|---|---|---|
| **1.** | 0.85<br>−0.64 | **2.** | $5.45<br>− 2.15 | **3.** | 1.69<br>−0.24 | **4.** | 4.29<br>−2.74 | **5.** | 0.75<br>−0.09 |
| **6.** | 24.3<br>−19.3 | **7.** | 4.7<br>−0.5 | **8.** | 0.9<br>−0.3 | **9.** | 7.7<br>−0.8 | **10.** | 0.26<br>−0.17 |
| **11.** | 4.2<br>−2.18 | **12.** | 0.9<br>−0.423 | **13.** | 9.15<br>−5.69 | **14.** | 0.33<br>−0.17 | **15.** | 2.06<br>−1.88 |
| **16.** | $5.00<br>− 1.25 | **17.** | 0.926<br>0.398 | **18.** | 0.55<br>−0.383 | **19.** | 4<br>−2.46 | **20.** | 88.3<br>−7.75 |
| **21.** | 9<br>−0.38 | **22.** | 8.425<br>−2.7 | **23.** | 48.6<br>−2.6 | **24.** | 3.784<br>−2.53 | **25.** | $5.98<br>− 5.09 |
| **26.** | 4.7<br>−2.845 | **27.** | 21.5<br>−11.96 | **28.** | 86.73<br>29.8 | **29.** | 0.489<br>−0.409 | **30.** | 8.4<br>−2.965 |

*Solve.*

**31.** Sylvia can stroke 12,789 characters per hour as a keypuncher. If the average stroke is 14,000 characters per hour, how much less does Sylvia do?

**32.** John spent $14.75 last week and $21.32 this week. How much more did he spend this week?

*Subtract.*

**33.** 26.78 − 0.84

**34.** 8.66 − 1.77

**35.** 89.51 − 31.24

**36.** 67.2 − 50.97

**37.** 31.65 − 20.07

**38.** 8.47 − 0.88

**39.** 9.9 − 0.95

**40.** 346.62 − 38.82

**41.** 34.9 − 22.11

**42.** 9.48 − 6.47

**43.** 47.36 − 18.36

**44.** 59.33 − 0.695

**45.** 51.94 − 9.99

**46.** 6.707 − 4.629

**47.** 2.235 − 1.273

**48.** 34.813 − 28.584

**49.** 7.502 − 0.4278

**50.** 3.219 − 2.774

**51.** 20.99 − 19.44

**52.** 9.194 − 2.177

Subtract 3.41 from 6.12.

First, align decimal points.

$$\begin{array}{r} 6.12 \\ -3.41 \\ \hline \end{array}$$

Then subtract as with whole numbers.

$$\begin{array}{r} {}^{5\,11} \\ 6.12 \\ -3.41 \\ \hline 2.71 \end{array}$$

*Subtract.*

**53.** 9.53 − 1.69

**54.** 2.78 − 1.49

**55.** 9.05 − 2.47

**56.** 22.07 − 13.69

**57.** 67.66 − 48.88

**58.** 12.065 − 1.058

**59.** 1.8 − 0.9

**60.** 71.81 − 61.96

**61.** 7.407 − 6.91

**62.** 3.16 − 0.76

**63.** 10.84 − 3.80

**64.** 79.3 − 3.93

**65.** 4 − 0.6

**66.** 7.46 − 5.7

**67.** 0.96 − 0.40

**68.** 6.51 − 0.8

**69.** 8.426 − 1.5

**70.** 11 − 0.141

*Solve.*

**71.** Before a 685.8 mile trip, the odometer reads 23,805.9. What is the reading after the trip?

**72.** A box of candy is purchased for $16.59. How much change is given from a $20 bill?

# Multiplying Decimals

Multiply by the number in each place-value position from least to greatest as with whole numbers. Then, answer the following to find out how to place the decimal point.

**A**  How many decimal places are in 6.243?

**B**  How many decimal places are in 0.85?

**C**  From this information, how can you determine the number of decimal places in the product?

6.243 _____ *3 decimal places*

× 0.85 _____ *2 decimal places*

31215

49944

5.30655 _____ *5 decimal places*

> **The number of decimal places in the product is the same as the sum of the decimal places in the factors.**

*For each of the following, multiply and find the number of decimal places in each factor and in the product.*

| | | | | |
|---|---|---|---|---|
| **1.** $$\begin{array}{r}\$1.45\\ \times\quad 0.6\\\hline\end{array}$$ | **2.** $$\begin{array}{r}0.8\\ \times\ 0.5\\\hline\end{array}$$ | **3.** $$\begin{array}{r}0.7\\ \times\ 0.7\\\hline\end{array}$$ | **4.** $$\begin{array}{r}24.8\\ \times\ 0.1\\\hline\end{array}$$ | **5.** $$\begin{array}{r}2.05\\ \times\ 0.9\\\hline\end{array}$$ |
| **6.** $$\begin{array}{r}0.67\\ \times\ 0.3\\\hline\end{array}$$ | **7.** $$\begin{array}{r}9.8\\ \times\ 0.02\\\hline\end{array}$$ | **8.** $$\begin{array}{r}0.7\\ \times\ 0.08\\\hline\end{array}$$ | **9.** $$\begin{array}{r}9.3\\ \times\ 0.2\\\hline\end{array}$$ | **10.** $$\begin{array}{r}10.1\\ \times\ 0.1\\\hline\end{array}$$ |
| **11.** $$\begin{array}{r}0.987\\ \times\quad 0.6\\\hline\end{array}$$ | **12.** $$\begin{array}{r}6.05\\ \times\ 0.21\\\hline\end{array}$$ | **13.** $$\begin{array}{r}1.064\\ \times\quad 0.9\\\hline\end{array}$$ | **14.** $$\begin{array}{r}9.06\\ \times\ 4.4\\\hline\end{array}$$ | **15.** $$\begin{array}{r}1.41\\ \times\ 0.34\\\hline\end{array}$$ |

*Multiply.*

| | | | | |
|---|---|---|---|---|
| **16.** $$\begin{array}{r}0.8\\ \times\ 8\\\hline\end{array}$$ | **17.** $$\begin{array}{r}23\\ \times\ 0.4\\\hline\end{array}$$ | **18.** $$\begin{array}{r}\$8.71\\ \times\quad 3\\\hline\end{array}$$ | **19.** $$\begin{array}{r}174\\ \times\ 0.07\\\hline\end{array}$$ | **20.** $$\begin{array}{r}0.819\\ \times\quad 3\\\hline\end{array}$$ |
| **21.** $$\begin{array}{r}0.2\\ \times\ 0.9\\\hline\end{array}$$ | **22.** $$\begin{array}{r}3.2\\ \times\ 0.1\\\hline\end{array}$$ | **23.** $$\begin{array}{r}44.4\\ \times\ 0.2\\\hline\end{array}$$ | **24.** $$\begin{array}{r}\$7.25\\ \times\quad 0.6\\\hline\end{array}$$ | **25.** $$\begin{array}{r}82.1\\ \times\ 0.07\\\hline\end{array}$$ |
| **26.** $$\begin{array}{r}16\\ \times\ 4.2\\\hline\end{array}$$ | **27.** $$\begin{array}{r}482\\ \times\ 1.5\\\hline\end{array}$$ | **28.** $$\begin{array}{r}0.479\\ \times\quad 0.2\\\hline\end{array}$$ | **29.** $$\begin{array}{r}4.629\\ \times\quad 0.9\\\hline\end{array}$$ | **30.** $$\begin{array}{r}3.48\\ \times\ 6.5\\\hline\end{array}$$ |
| **31.** $$\begin{array}{r}16.8\\ \times\ 0.003\\\hline\end{array}$$ | **32.** $$\begin{array}{r}0.429\\ \times\quad 3.4\\\hline\end{array}$$ | **33.** $$\begin{array}{r}16.73\\ \times\ 0.019\\\hline\end{array}$$ | **34.** $$\begin{array}{r}\$6570\\ \times\quad 0.82\\\hline\end{array}$$ | **35.** $$\begin{array}{r}825\\ \times\ 0.7\\\hline\end{array}$$ |
| **36.** $$\begin{array}{r}\$3.75\\ \times\quad 2.7\\\hline\end{array}$$ | **37.** $$\begin{array}{r}8.07\\ \times\ 6.53\\\hline\end{array}$$ | **38.** $$\begin{array}{r}5.7\\ \times\ 0.36\\\hline\end{array}$$ | **39.** $$\begin{array}{r}9.108\\ \times\quad 5.2\\\hline\end{array}$$ | **40.** $$\begin{array}{r}\$27.13\\ \times\quad 4.8\\\hline\end{array}$$ |

| 41. | 4.76<br>× 1.4 | 42. | 44.5<br>× 0.66 | 43. | 0.352<br>× 2.5 | 44. | 5.28<br>× 0.92 | 45. | 71.6<br>× 2.2 |
|---|---|---|---|---|---|---|---|---|---|
| 46. | 4.623<br>× 0.807 | 47. | $706.85<br>× 1.42 | 48. | 9.7302<br>× 0.604 | 49. | 85.71<br>× 3.09 | 50. | $6.55<br>× 4.2 |
| 51. | 0.531<br>× 2.73 | 52. | 8.294<br>× 64.5 | 53. | 0.724<br>× 0.42 | 54. | 1.386<br>× 2.5 | 55. | 86.06<br>× 0.118 |
| 56. | $37.25<br>× 23.4 | 57. | 68.052<br>× 4.73 | 58. | 0.7428<br>× 3.25 | 59. | 21.096<br>× 4.62 | 60. | 402.6<br>× 0.819 |
| 61. | 73.408<br>× 15.5 | 62. | 5.0023<br>× 2.12 | 63. | 804.03<br>× 0.275 | 64. | 1.383<br>× 8.61 | 65. | 0.2828<br>× 0.733 |

*Solve.*

66. The trans-Alaska pipeline delivers up to 2 million barrels of oil per day. The pipeline is 800 miles long. At most, how many barrels can be delivered in 4.5 days?

67. Sean sells 15 gerbils to a pet store. they pay him 75¢ per gerbil. How much is he paid?

68. Mr. Wu paid $19.50 for 13 gallons of gasoline. His car gets 26.7 miles per gallon. How far can he drive his car on 13 gallons of gasoline?

69. The mayor's campaign expenses show that 15,000 bumper stickers are purchased at $0.0675 each. How much is the total cost for bumper stickers?

70. Mrs. Phillips paid $12.54 for 13.2 gallons of gasoline. Her car gets 22 miles per gallon. How far does she drive on 13.2 gallons of gasoline?

71. Gustav wants 15 tropical fish for his aquarium. The fish cost $2.98 each. How much does he need to buy the fish?

# Division of Decimals

To divide a decimal by a whole number, divide in each place-value position as with whole numbers.

Divide 99.25 by 25.

| First, divide the ones. | Divide the tenths. | Divide the hundredths. |
|---|---|---|
| 99 ones ÷ 25 | 242 tenths ÷ 25 | 175 ÷ 25 |

```
      3.
25)99.25
 - 75
```

```
      3.9
25)99.25
 - 75
   24 2
   22 5
```

```
      3.97
25)99.25
 - 75
   24 2
 - 22 5
    1 75
  - 1 75
```

*Write a 3 in the ones place of the quotient and place the decimal point.*

*Write a 9 in the tenths place in the quotient.*

*Write a 7 in the hundredths place in the quotient.*

*Copy and place the decimal point in the quotient.*

| | | | | |
|---|---|---|---|---|
| 0 042 | 0 021 | 0 82 | 0 7 | 16 1 |
| 1. 2)0.084 | 2. 4)0.084 | 3. 9)7.38 | 4. 28)19.6 | 5. 6)96.6 |

A lesser number can be divided by a greater number using decimals. Place the decimal point in the quotient. Annex zeros to the dividend in order to continue the division process. Round to the thousandths place.

| Place the decimal point in the quotient. | Annex zeros. | Divide as with whole numbers. | Thus, |
|---|---|---|---|
| 0.<br>7)5 | 0.7<br>7)5.00<br> - 4 9<br>   10 | 0.7142—<br>7)5.0000<br> - 4 9<br>   10<br> -  7<br>   30<br> - 28<br>   20<br> - 14 | 5 ÷ 7 ≈ 0.714 |

*Divide.*

6. 7)42          7. 7)4.2          8. 7)0.42          9. 7)0.042          10. 7)42.07

11. 7)3.5        12. 9)0.63        13. 4)0.84        14. 6)5.4            15. 7)9.1

16. 6)1.08       17. 9)12.6        18. 5)1.95        19. 8)0.584         20. 22)10.12

21. 14)0.378     22. 9)0.414       23. 39)25.74      24. 45)7.65         25. 26)5.59

*Divide.*

**26.** $25\overline{)4}$  **27.** $8\overline{)2}$  **28.** $5\overline{)2}$  **29.** $5\overline{)1}$  **30.** $6\overline{)3}$

**31.** $8\overline{)6}$  **32.** $4\overline{)3}$  **33.** $4\overline{)1}$  **34.** $50\overline{)40}$  **35.** $12\overline{)9}$

**36.** $16\overline{)13}$  **37.** $45\overline{)18}$  **38.** $6\overline{)5}$  **39.** $9\overline{)7}$  **40.** $12\overline{)7}$

**41.** $96\overline{)36}$  **42.** $3\overline{)2}$  **43.** $36\overline{)27}$  **44.** $25\overline{)19}$  **45.** $8\overline{)7}$

**46.** $64\overline{)21}$  **47.** $10\overline{)3}$  **48.** $7\overline{)3}$  **49.** $24\overline{)13}$  **50.** $13\overline{)8}$

**51.** $25\overline{)40}$  **52.** $4\overline{)946}$  **53.** $28\overline{)63}$  **54.** $60\overline{)444}$  **55.** $40\overline{)100}$

**56.** $20\overline{)117}$  **57.** $80\overline{)372}$  **58.** $400\overline{)552}$  **59.** $88\overline{)44}$  **60.** $25\overline{)15}$

**61.** $550\overline{)209}$  **62.** $80\overline{)52}$  **63.** $950\overline{)7315}$

# Decimal Divisors

Change divisors that are decimals to whole numbers before performing the division operation.

As shown at the right, multiplying *both* the divisor and dividend by a power of 10 does not change the quotient.

Divide 121.8 by 0.06.

Multiply both the divisor and dividend by a power of 10 so that the divisor is a whole number. Then divide as with whole numbers.

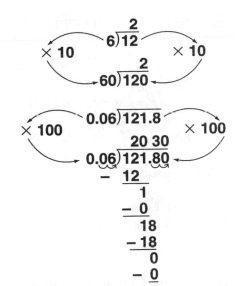

*Copy and place the decimal point in the quotient.*

$$
\begin{array}{ccccc}
0\ 22 & 3 & 0\ 05 & 70 & 90
\end{array}
$$

1.  6.1)1.342   2.  0.4)1.2   3.  7.6)0.380   4.  0.206)14.42   5.  0.56)50.4

*Divide.*

6.  0.7)0.49   7.  0.13)0.52   8.  0.4)7.2   9.  0.3)3.9   10.  0.8)6.4

11.  2.3)1.61   12.  0.21)0.168   13.  0.5)11.5   14.  1.6)4.8   15.  6.1)1.342

16.  0.003)0.018   17.  0.106)0.04876   18.  0.9)2.34   19.  7.9)40.29

20.  4.7)4.982   21.  0.06)5.34   22.  0.48)0.6   23.  3.5)42   24.  4.73)2379.19

25.  0.08)6.24   26.  0.52)0.832   27.  2.9)24.07   28.  0.618)139.05

29.  0.6)31.2   30.  0.94)7.99   31.  6.7)21.306   32.  0.08)0.4152   33.  1.95)0.0312

*Solve.*

34. Sonia buys enough tile to cover 112.5 sq ft. The width of the floor is 12.5 ft. What length does the tile cover?

35. Ticket sales for the game total $3860. If each ticket costs $2.50, how many tickets are sold?

36. There is $65 in the budget to buy new basketballs. Each one costs $14.95. How many basketballs are bought?

37. A bus gets 12.4 miles per gallon. How many miles does the bus travel on 53.7 gallons of gasoline?

# Area of Circle

The area of a parallelogram is found by multiplying base and height. If a circle is cut apart, as shown below, this same idea can be used.

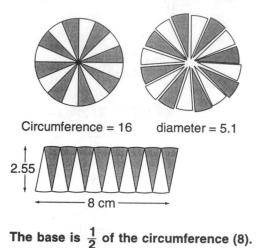

Circumference = 16     diameter = 5.1

2.55

8 cm

| | |
|---|---|
| $A = b \times h$ | *Area formula of a parallelogram.* |
| $A = \left(\frac{1}{2} \times C\right) \times h$ | *You can replace b by $\frac{1}{2} \times C$. Why?* |
| $A = \frac{1}{2} \times C \times r$ | *Why can you replace h by r?* |
| $A = \frac{1}{2} \times (\pi \times d) \times r$ | *Remember, $C = \pi \times d$.* |
| $A = \frac{1}{2} \times (\pi \times 2 \times r) \times r$ | *2 × the radius = diameter* |
| $A = \pi \times r \times r$ | *How is this obtained?* |
| $A = \pi \times r^2$ | *$r^2$ means $r \times r$* |

The base is $\frac{1}{2}$ of the circumference (8).

The height is $\frac{1}{2}$ of the diameter, or the radius (2.55)

$A = b \times h$     $A = 8 \times 2.55$     $A = 20.4$

Therefore, the area of a circle can be found by multiplying $\pi$ (3.14) by the radius squared $(A = \pi \times r^2)$.

Miles MacLean is making a circular cake. He needs to know the area of the circle so he can make frosting for the cake

$A = \pi \times r \times r$
$A \approx 3.14 \times 10 \times 10$
$A \approx 314$

Miles needs 314 sq cm of frosting.

*Find the area of each circle for which the measure of the radius is given. Use 3.14 for $\pi$. **Round answers to the nearest hundredth.***

| | | |
|---|---|---|
| **1.** 6 m | **2.** 1.7 ft | **3.** 4 in. |
| **4.** 0.2 cm | **5.** 1.3 mm | **6.** 4.4 m |
| **7.** 9.5 m | **8.** 2.9 mi | **9.** 1.3 cm |
| **10.** 17 ft | **11.** 12 km | **12.** 66 yd |
| **13.** 18 m | **14.** 3.3 ft | **15.** 7.9 dm |

# Fractions to Decimals

$$\frac{1}{4} \rightarrow 1 \div 4 \rightarrow 4\overline{)1.00}$$

Notice that the division ends or terminates. A decimal such as 0.25 is called a **terminating decimal**.

In certain division problems, one or more of the digits in the quotient will repeat.

0.833 or 0.8$\overline{3}$
6$\overline{)5.000}$
$-48$
  20
$-18$
  20
$-18$
   2

*No matter how far you compute, the 3 repeats in the quotient.*

72.727 or 72.$\overline{72}$
0.11$\overline{)8.00.000}$
$-77$
  30
$-22$
  80
$-77$
  30
$-22$
  80
$-77$
   3

*What numbers continue to repeat in the quotient?*

*In division problems like these, compute until you are sure the pattern repeats.*

Repeating decimals are identified with bar notation and indicate that the digits repeat indefinitely.

$$0.\overline{6} = 0.66666\ldots \qquad 0.8\overline{3} = 0.83333\ldots \qquad 72.\overline{72} = 72.727272\ldots$$

Decimals such as $0.\overline{6}$, $0.8\overline{3}$, and $72.\overline{72}$ are called **repeating decimals**.

*Copy and complete each of the following using bar notation.*

1. 0.373737...
2. 0.44444...
3. 0.73333...
4. 2.9612612...
5. 0.010101...
6. 0.92555...

*Divide. Use bar notation in the quotient.*

7. 0.6$\overline{)1.3}$
8. 9$\overline{)572}$
9. 0.6$\overline{)0.5}$
10. 11$\overline{)600}$
11. 24$\overline{)500}$
12. 3$\overline{)40}$
13. 0.24$\overline{)19}$
14. 0.3$\overline{)0.08}$

Divide. Compare to the thousandth place or until you are sure the pattern repeats.

**15.** $25\overline{)6.25}$    **16.** $6.1\overline{)0.2623}$    **17.** $0.57\overline{)3.42}$    **18.** $6\overline{)4}$    **19.** $8\overline{)5}$

**20.** $15\overline{)11}$    **21.** $0.4\overline{)1.4}$    **22.** $12\overline{)8}$    **23.** $3.8\overline{)20.52}$    **24.** $11\overline{)1.7}$

**25.** $999\overline{)8}$    **26.** $0.72\overline{)0.0648}$   **27.** $11\overline{)9}$    **28.** $59\overline{)159.3}$    **29.** $18\overline{)15}$

Change each fraction to a decimal. Use a bar to show a repeating decimal.

**30.** $\dfrac{7}{9}$    **31.** $\dfrac{7}{8}$    **32.** $\dfrac{5}{18}$    **33.** $\dfrac{7}{16}$    **34.** $\dfrac{5}{12}$

**35.** $\dfrac{13}{25}$    **36.** $\dfrac{11}{15}$    **37.** $\dfrac{1}{6}$    **38.** $\dfrac{3}{5}$    **39.** $\dfrac{2}{3}$

**40.** $\dfrac{6}{11}$    **41.** $\dfrac{5}{9}$    **42.** $\dfrac{9}{10}$    **43.** $\dfrac{1}{4}$    **44.** $\dfrac{19}{30}$

**45.** $\dfrac{22}{45}$    **46.** $\dfrac{17}{20}$    **47.** $\dfrac{21}{40}$    **48.** $\dfrac{4}{7}$    **49.** $\dfrac{31}{60}$

# ANSWERS and SOLUTIONS for DECIMALS

## Decimal Place Value (pages 39–40)

1. one  2. thousand  3. thousandth  4. hundred  5. tenth  6. ten  7. tenth  8. hundredth  9. hundred
10. ten thousandth  11. ten  12. thousandth  13. thousand  14. one  15. 7  16. 4000  17. 0.006  18. 500
19. 0.3  20. 20  21. 7000  22. 0.05  23. 0.0006  24. 8  25. 30  26. 0.004  27. 0.00001  28. 200
29. 200  30. 9  31. 0.1  32. 0.005  33. 60  34. 0.0003  35. 4000  36. 0.08  37. tenth  38. thousandth
39. thousandth  40. tenth  41. thousandth  42. hundredth  43. hundred  44. tenth  45. ten  46. thousandth
47. hundredth  48. hundredth  49. one  50. thousandth  51. hundredth  52. one  53. ten  54. tenth  55. tenth
56. tenth  57. tenth  58. thousandth  59. hundredth  60. hundred  61. hundred  62. hundredth  63. tenth
64. twenty-four and seventy-one hundredths  65. seventeen and ninety-four hundredths
66. thirty-three and twenty-five hundredths  67. eleven and three hundred ninety-four thousandths
68. thirty-seven and ninety-eight hundredths  69. seven hundred four thousandths
70. eighty-two and nine hundred fifty-one thousandths  71. fourteen and three hundred six thousandths
72. forty-eight and eight thousand one hundred ninety-two ten thousandths.

73. thirteen and three hundred twenty-five thousandths  74. seven and $\frac{65}{100}$  75. fifteen and $\frac{00}{100}$

76. sixty-three and $\frac{59}{100}$  77. forty-five and $\frac{07}{100}$  78. twenty-one and $\frac{08}{100}$

79. one hundred twenty-five and $\frac{14}{100}$  80. two and $\frac{98}{100}$  81. three hundred seventy-six and $\frac{23}{100}$

82. fifty and $\frac{00}{100}$  83. two thousand nine hundred forty eight and $\frac{31}{100}$  84. eighty four and $\frac{72}{100}$

## Comparing Decimals (page 41)

1. $400 = 400$, true  2. $30 \neq 03$, false  3. $10 \neq 01$, false  4. $2002 \neq 2020$, false  5. $180 = 180$ true
6. $710 = 710$, true  7. $3330 \neq 3333$, false  8. $1400 = 1400$, true  9. $470 = 470$, true  10. $08 < 80$, true
11. $7 \not> 50$, false  12. $05 \neq 55$, false  13. $50 \neq 500$, false  14. $9 > 7$, true  15. $6000 < 6001$, true
16. $10002 \not> 10020$, false  17. $8050 \not< 8002$, false  18. $3 \not> 30$, false  19. false  20. true  21. true  22. true
23. false  24. true  25. true  26. true  27. true  28. false  29. true  30. true
31. 2222.37, 22.4, 22.7, 23.4, 23.7, 24, 24.6  32. $0.9 = 0.90$  33. $0.6 < 0.7$  34. $0.904 < 0.909$  35. $0.09 < 0.90$
36. $0.06 < 0.60$  37. $2.03 < 3.02$  38. $0.49 > 0.44$  39. $0.84 = 0.840$  40. $6.216 > 6.215$

## Rounding Decimals (page 42)

| | | | | | | | |
|---|---|---|---|---|---|---|---|
| 1. 33.7 | 2. 77.3 | 3. 5.5 | 4. 49.7 | 5. 28.5 | 6. 6.6 | 7. 21.4 | 8. 13.6 |
| 9. 4.9 | 10. 10.3 | 11. 9 | 12. 8.8 | 13. 8.76 | 14. 8.565 | 15. 0.6 | 16. 1.79 |
| 17. 5.81 | 18. 39.13 | 19. 1.86 | 20. 0.87 | 21. 0.98 | 22. 8.22 | 23. 0.59 | 24. 21.28 |
| 25. 3.97 | 26. 0.05 | 27. 6.3 | 28. 23 | 29. 6.9 | 30. 0.009 | 31. 14.1 | 32. 6.011 |
| 33. $6 | 34. $9.27 | 35. 0.31 | 36. 6 | 37. 8 | 38. 1 | 39. 8 | 40. 3 |
| 41. $8 | 42. $2 | 43. 9 | 44. 10 | 45. 7 | 46. 32.1 | 47. 0.3 | 48. 0.9 |
| 49. 0.1 | 50. 0.2 | 51. 6.3 | 52. 0.2 | 53. 45.0 | 54. 1.0 | 55. 0.8 | 56. 7.65 |
| 57. 3.44 | 58. $8.05 | 59. 66.07 | 60. 0.16 | 61. 9.07 | 62. 0.06 | 63. $8.25 | 64. 0.05 |
| 65. $3.10 | 66. 0.007 | 67. 0.003 | 68. 0.089 | 69. 5.107 | 70. 49.014 | 71. 6.970 | 72. $0.610 |
| 73. $0.325 | 74. 7.661 | 75. $0.004 | | | | | |

## Addition of Decimals (pages 43–44)

| | | | | | | | |
|---|---|---|---|---|---|---|---|
| 1. 0.9 | 2. 7.3 | 3. 9.4 | 4. 8.1 | 5. 8.5 | 6. 11.8 | 7. 8.9 | 8. 4.8 |
| 9. 8.8 | 10. 0.56 | 11. 76.7 | 12. 7.883 | 13. $59.97 | 14. 7.84 | 15. 8.895 | 16. 10.679 |
| 17. 87.888 | 18. 29.985 | 19. 47.884 | 20. 83.589 | 21. 12.2 | 22. $0.44 | 23. 4.4 | 24. 3.6 |
| 25. 5.55 | 26. 1.74 | 27. 8.37 | 28. 78.2 | 29. 47.07 | 30. 5.98 | 31. 6.15 | 32. $8.35 |
| 33. 19.4 | 34. 16.20 | 35. 8.054 | 36. 1.839 | 37. 972.1 | 38. 9.49 | 39. 14.13 | 40. 14.24 |

| 41. 108.3 | 42. $13.47 | 43. 1.020 | 44. 142.0 | 45. 7.362 | 46. 57.314 | 47. 69.236 | 48. 150.54 |
|---|---|---|---|---|---|---|---|
| 49. 70.538 | 50. 66.213 | 51. 52.196 | 52. 79.529 | 53. 97.454 | 54. 113.18 | 55. 113.87 | |

56.
```
   55.18
 +16.93
  72.11
```

57.
```
  2.15
 +0.22
  2.37
```

58.
```
   0.5
   3.9
 + 8.3
  12.7
```

59.
```
  12.75
 +0.25
  13.00
```

60.
```
   1.90
 +2.01
   3.91
```

61.
```
   48.720
    6.195
 + 0.060
   54.975
```

62.
```
   98.08
 +16.93
  115.01
```

63.
```
    1.090
    3.800
   17.320
 +0.758
   22.968
```

64.
```
   2.854
   7.600
 +3.420
  13.874
```

65.
```
  23.95
  16.09
  14.00
 +1.82
  55.86
```

66.
```
  $15.70   F.I.C.A. tax
   52.89   withholding tax
    3.42   local tax
 + 8.65   deductions
  $80.66   total deductions
```

67.
```
   2.46   cm   present size
 +0.09   cm   amount worn away
   2.55   cm   original size
```

## Subtraction of Decimals (pages 45–46)

| 1. 0.21 | 2. $3.30 | 3. 1.45 | 4. 1.55 | 5. 0.66 | 6. 5.0 | 7. 4.2 |
|---|---|---|---|---|---|---|
| 8. 0.6 | 9. 6.9 | 10. 0.09 | 11. 2.02 | 12. 0.477 | 13. 3.46 | 14. 0.16 |
| 15. 0.18 | 16. $3.75 | 17. 0.528 | 18. 0.167 | 19. 1.54 | 20. 80.55 | 21. 8.62 |
| 22. 5.725 | 23. 46.0 | 24. 1.254 | 25. $0.89 | 26. 1.855 | 27. 9.54 | 28. 56.93 |
| 29. 0.080 | 30. 5.435 | | | | | |

31.
```
   14,000   average character strokes
 -12,789   Sylvia character strokes
    1211   fewer characters stroked
```

32.
```
  $21.32   this week
 -14.75   last week
 $ 6.57   more spent
```

| 33. 25.94 | 34. 6.89 | 35. 58.27 | 36. 16.23 | 37. 11.58 | 38. 7.59 | 39. 8.95 |
|---|---|---|---|---|---|---|
| 40. 307.80 | 41. 12.79 | 42. 3.01 | 43. 29.00 | 44. 58.635 | 45. 41.95 | 46. 2.078 |
| 47. 0.962 | 48. 6.229 | 49. 7.0742 | 50. 0.445 | 51. 1.55 | 52. 7.017 | |

53.
```
   9.53
 -1.69
   7.84
```

54.
```
   2.78
 -1.49
   1.29
```

55.
```
   9.05
 -2.47
   6.58
```

56.
```
  22.07
 -13.69
   8.38
```

57.
```
  67.66
 -48.88
  18.78
```

58.
```
  12.065
 -1.058
  11.007
```

59.
```
   1.8
 -0.9
   0.9
```

60.
```
   71.81
 -61.96
   9.85
```

61.
```
   7.407
 -6.910
   0.497
```

62.
```
   3.16
 -0.76
   2.40
```

63.
```
  10.84
 -3.80
   7.04
```

64.
```
  79.30
 -3.93
  75.37
```

65.
```
   4.0
 -0.6
   3.4
```

66.
```
   7.46
 -5.70
   1.76
```

67.
```
   0.96
 -0.40
   0.56
```

68.
```
   6.51
 - 0.80
   5.71
```

69.
```
   8.426
 -1.5
   6.926
```

70.
```
  11.000
 -0.141
  10.859
```

71.
```
  23805.9   odometer beginning
 + 685.8   length of trip
  24491.7   odometer after trip
```

72.
```
  $20.00   paid in cash
 -16.59   purchase price
 $ 3.41   change
```

## Multiplying Decimals (pages 47–48)

1.
```
 $ 1.45   2 decimal places
 ×  0.6   1 decimal place
 $0.870   3 decimal places
```

2.
```
    0.8   1 decimal place
 ×0.5   1 decimal place
   0.40   2 decimal places
```

3.
```
    0.7   1 decimal place
 ×0.7   1 decimal place
   0.49   2 decimal places
```

4.
```
   24.8   1 decimal place
 ×0.1   1 decimal place
   2.48   2 decimal places
```

5.
```
   2.05   2 decimal places
 × 0.9   1 decimal place
   1.845   3 decimal places
```

6.
```
   0.67   2 decimal places
 × 0.3   1 decimal place
   0.201   3 decimal places
```

7.
```
    9.8   1 decimal place
 ×0.02   2 decimal places
   0.196   3 decimal places
```

8.
```
    0.7   1 decimal place
 ×0.08   2 decimal places
   0.056   3 decimal places
```

9.
```
    9.3   1 decimal place
 ×0.2   1 decimal place
   1.86   2 decimal places
```

**10.**
```
  10.1   1 decimal place
×  0.1   1 decimal place
 1.01    2 decimal places
```

**11.**
```
  0.987   3 decimal places
×   0.6   1 decimal place
 0.5922   4 decimal places
```

**12.**
```
   6.05   2 decimal places
×  0.21   2 decimal places
    605
   1210
 1.2705   4 decimal places
```

**13.**
```
  1.064   3 decimal places
×   0.9   1 decimal place
 0.9576   4 decimal places
```

**14.**
```
   9.06   2 decimal places
×   4.4   1 decimal place
   3624
  3624
 39.864   3 decimal places
```

**15.**
```
   1.41   2 decimal places
× 0.34    2 decimal places
    564
   423
 0.4794   4 decimal places
```

**16.**
```
  0.8
×   8
  6.4
```

**17.**
```
    23
×  0.4
   9.2
```

**18.**
```
 $8.71
×    3
 26.13
```

**19.**
```
   174
× 0.07
 12.18
```

**20.**
```
  0.819
×     3
  2.457
```

**21.**
```
   0.2
× 0.9
 0.18
```

**22.**
```
   3.2
× 0.1
 0.32
```

**23.**
```
  44.4
× 0.2
 8.88
```

**24.**
```
 $7.25
×   0.6
 $4.350
```

**25.**
```
   82.1
× 0.07
 5.747
```

**26.**
```
    16
× 4.2
   32
  64
 67.2
```

**27.**
```
   482
×  1.5
  2410
  482
 723.0
```

**28.**
```
  0.479
×   0.2
 0.0958
```

**29.**
```
  4.629
×   0.9
 4.1661
```

**30.**
```
   3.48
×   6.5
  1740
 2088
 22.620
```

**31.**
```
   16.8
× 0.003
 0.0504
```

**32.**
```
  0.429
×   3.4
  1716
 1287
 1.4586
```

**33.**
```
  16.73
× 0.019
 15057
 1673
 0.31787
```

**34.**
```
 $6570
×  0.82
 13140
 52560
 $5387.40
```

**35.**
```
   825
× 0.7
 577.5
```

**36.**
```
 $3.75
×   2.7
  2625
 750
 $10.125
```

**37.**
```
   8.07
× 6.53
  2421
 4035
 4842
 52.6971
```

**38.**
```
    5.7
× 0.36
  342
 171
 2.052
```

**39.**
```
  9.108
×   5.2
 18216
 45540
 47.3616
```

**40.**
```
 $27.13
×    4.8
  21704
 10852
 $130.224
```

**41.**
```
   4.76
×  1.4
  1904
 476
 6.664
```

**42.**
```
   44.5
× 0.66
  2670
 2670
 29.370
```

**43.**
```
  0.352
×   2.5
  1760
 704
 0.8800
```

**44.**
```
   5.28
× 0.92
  1056
 4752
 4.8576
```

**45.**
```
   71.6
×  2.2
  1432
 1432
 157.52
```

**46.**
```
  4.623
× 0.807
  32361
 369840
 3.730761
```

**47.**
```
 $706.85
×    1.42
  141370
 282740
 70685
 $1003.7270
```

**48.**
```
  9.7302
×  0.604
  389208
 583812
 5.8770408
```

**49.**
```
  85.71
×  3.09
  77139
 25713
 264.8439
```

**50.**
```
 $6.55
×   4.2
  1310
 2620
 $27.510
```

**51.**
```
  0.531
×  2.73
  1593
 3717
 1062
 1.44963
```

**52.**
```
  8.294
×  64.5
  41470
 33176
 49764
 534.9630
```

**53.**
```
  0.724
×  0.42
  1448
 2896
 0.30408
```

**54.**
```
  1.386
×   2.5
  6930
 2772
 3.4650
```

**55.**
```
  86.06
× 0.118
  68848
 8606
 8606
 10.15508
```

**56.**
```
 $37.25
×   23.4
  14900
 11175
 7450
 $871.650
```

**57.**
```
  68.052
×   4.73
  204156
 476364
 272208
 321.88596
```

**58.**
```
  0.7428
×   3.25
  37140
 14856
 22284
 2.414100
```

**59.**
```
  21.096
×   4.62
  42192
 126576
 84384
 97.46352
```

**60.**
```
   402.6
× 0.819
  36234
 4026
 32208
 329.7294
```

**61.**
```
  73.408
×   15.5
  367040
 367040
 73408
 1137.8240
```

**62.**
```
  5.0023
×   2.12
  100046
 50023
 100046
 10.604876
```

**63.**
```
  804.03
× 0.275
  402015
 562821
 160806
 221.10825
```

**64.**
```
  1.383
×  8.61
  1383
 8298
 11064
 11.90763
```

**65.**
```
  0.2828
× 0.733
  8484
 8484
 19796
 .2072924
```

**66.**
```
    4.5   days
  ×   2   million barrels per day
  9.0   million barrels
```

**67.**
```
  $0.75   price per gerbil
  ×   15   number of gerbils
     375
      75
  $11.25   total received
```

**68.**
```
   26.7   miles per gallon
  ×   13   gallons
    80 1
   267
   347.1   miles
```

**69.**
```
  $0.0675   cost per sticker
  ×   15000   number of stickers
    3375000
     675
  $1012.5000   cost for stickers
```

**70.**
```
   13.2   gallons
  ×   22   miles per gallon
    26 4
   264
   290.4   miles
```

**71.**
```
  $2.98   price per fish
  ×   15   number of fish
    1490
     298
  $44.70   total cost
```

## Division of Decimals (pages 49–50)

**1.** 0.042    **2.** 0.021    **3.** 0.82    **4.** 0.7    **5.** 16.1

**6.**
```
      6
  7)42
    42
```

**7.**
```
     0.6
  7)4.2
    4 2
```

**8.**
```
     0.06
  7)0.42
    0 42
```

**9.**
```
     0.006
  7)0.042
    0 042
```

**10.**
```
     6.01
  7)42.07
    42
      0
      0
       7
       7
```

**11.**
```
     0.5
  7)3.5
    3 5
```

**12.**
```
     0.07
  9)0.63
    0
     6
     0
    63
    63
```

**13.**
```
     0.21
  4)0.84
    0
     8
     8
      4
      4
```

**14.**
```
     0.9
  6)5.4
    5 4
```

**15.**
```
     1.3
  7)9.1
    7
    2 1
    2 1
```

**16.**
```
     0.18
  6)1.08
    6
    48
    48
```

**17.**
```
     1.4
  9)12.6
    9
    3 6
    3 6
```

**18.**
```
     0.39
  5)1.95
    1 5
     45
     45
```

**19.**
```
     0.073
  8)0.584
    0
     5
     0
    58
    56
     24
     24
```

**20.**
```
     00.46
  22)10.12
     0
     10
      0
     10 1
      8 8
      1 32
      1 32
```

**21.**
```
      0.027
  14)0.378
     0
      3
      0
     37
     28
      98
      98
```

**22.**
```
     0.046
  9)0.414
    0
     4
     0
    41
    36
     54
     54
```

**23.**
```
      0.66
  39)25.74
     23 4
      2 34
      2 34
```

**24.**
```
     0.17
  45)7.65
     4 5
     3 15
     3 15
```

**25.**
```
     0.215
  26)5.590
     0
     5 5
     5 2
      39
      26
      130
      130
```

**26.**
```
     0.16
  25)4.00
     2 5
     1 50
     1 50
```

**27.**
```
     0.25
  8)2.00
    1 6
     40
     40
```

**28.**
```
     0.4
  5)2.0
    2 0
```

**29.**
```
     0.2
  5)1.0
    1 0
```

**30.**
```
     0.5
  6)3.0
    3 0
```

**31.**
```
     0.75
  8)6.00
    5 6
     40
     40
```

**32.**
```
     0.75
  4)3.00
    2 8
     20
     20
```

**33.**
```
     0.25
  4)1.00
    8
    20
    20
```

**34.**
```
     0.8
  50)40.0
     40 0
```

**35.**
```
     0.75
  12)9.00
     8 4
      60
      60
```

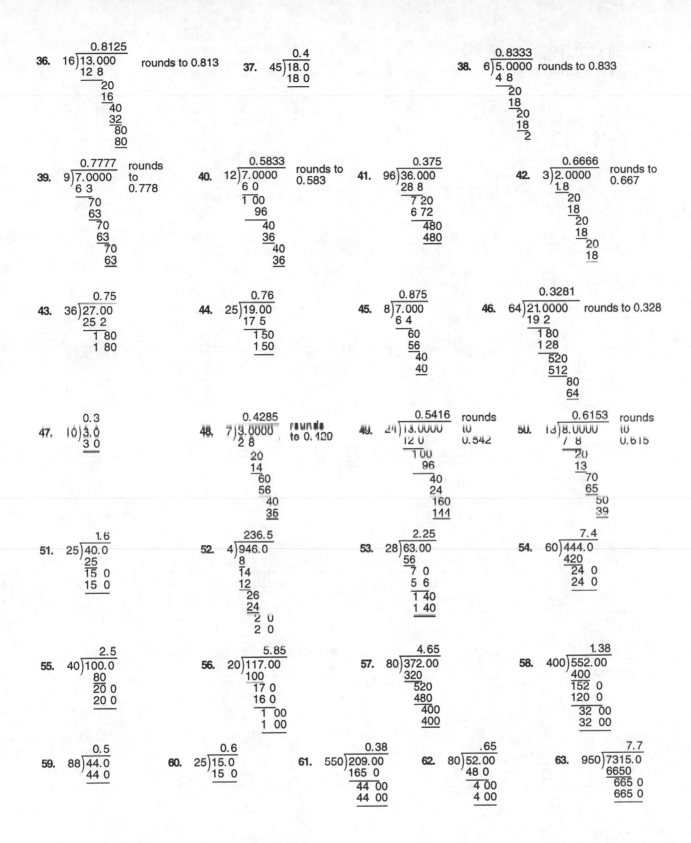

**36.**
```
      0.8125
 16)13.000      rounds to 0.813
    12 8
       20
       16
       40
       32
       80
       80
```

**37.**
```
      0.4
 45)18.0
    18 0
```

**38.**
```
     0.8333
 6)5.0000      rounds to 0.833
   4 8
     20
     18
     20
     18
      2
```

**39.**
```
     0.7777    rounds
 9)7.0000      to
   6 3         0.778
     70
     63
     70
     63
     70
     63
```

**40.**
```
      0.5833    rounds to
 12)7.0000      0.583
    6 0
    1 00
      96
      40
      36
      40
      36
```

**41.**
```
      0.375
 96)36.000
    28 8
     7 20
     6 72
       480
       480
```

**42.**
```
     0.6666     rounds to
 3)2.0000       0.667
   1.8
     20
     18
     20
     18
     20
     18
```

**43.**
```
      0.75
 36)27.00
    25 2
     1 80
     1 80
```

**44.**
```
      0.76
 25)19.00
    17 5
     1 50
     1 50
```

**45.**
```
     0.875
 8)7.000
   6 4
     60
     56
     40
     40
```

**46.**
```
      0.3281
 64)21.0000    rounds to 0.328
    19 2
     1 80
     1 28
       520
       512
        80
        64
```

**47.**
```
     0.3
 10)3.0
    3 0
```

**48.**
```
     0.4285    rounds
 7)3.0000      to 0.429
   2 8
     20
     14
     60
     56
     40
     35
```

**49.**
```
      0.5416    rounds
 24)13.0000     to
    12 0        0.542
     1 00
       96
       40
       24
       160
       144
```

**50.**
```
      0.6153    rounds
 13)8.0000      to
    7 8          0.615
      20
      13
      70
      65
      50
      39
```

**51.**
```
      1.6
 25)40.0
    25
    15 0
    15 0
```

**52.**
```
     236.5
 4)946.0
   8
   14
   12
    26
    24
     2 0
     2 0
```

**53.**
```
      2.25
 28)63.00
    56
     7 0
     5 6
     1 40
     1 40
```

**54.**
```
      7.4
 60)444.0
    420
     24 0
     24 0
```

**55.**
```
      2.5
 40)100.0
    80
    20 0
    20 0
```

**56.**
```
      5.85
 20)117.00
    100
     17 0
     16 0
      1 00
      1 00
```

**57.**
```
      4.65
 80)372.00
    320
     520
     480
      400
      400
```

**58.**
```
       1.38
 400)552.00
     400
     152 0
     120 0
      32 00
      32 00
```

**59.**
```
      0.5
 88)44.0
    44 0
```

**60.**
```
      0.6
 25)15.0
    15 0
```

**61.**
```
       0.38
 550)209.00
     165 0
      44 00
      44 00
```

**62.**
```
      .65
 80)52.00
    48 0
     4 00
     4 00
```

**63.**
```
       7.7
 950)7315.0
     6650
      665 0
      665 0
```

59

## Decimals Divisors (page 51)

**1.** 0.22  **2.** 3  **3.** 0.05  **4.** 70  **5.** 90

**6.**
```
       0.7
 0.7)0.4 9
     4 9
```

**7.**
```
        4.
 0.13)0.52
      52
```

**8.**
```
      18.
 0.4)7.2
     4
     3 2
     3 2
```

**9.**
```
      13.
 0.3)3.9
     3
      9
      9
```

**10.**
```
      8.
 0.8)6.4
     6 4
```

**11.**
```
       0.7
 2.3)1.61
     1 61
```

**12.**
```
        0.8
 0.21)0.16 8
      16 8
```

**13.**
```
       23.
 0.5)11.5
     10
      1 5
      1 5
```

**14.**
```
      3.
 1.6)4.8
     4 8
```

**15.**
```
        0.22
 6.1)1.3 42
     1 22
       1 22
       1 22
```

**16.**
```
         6.
 0.003)0.018
       18
```

**17.**
```
         0.46
 0.106)0.048 76
       42 4
        6 36
        6 36
```

**18.**
```
      2.6
 0.9)2.3 4
     1 8
      5 4
      5 4
```

**19.**
```
       5.1
 7.9)40.2 9
     39 5
       7 9
       7 9
```

**20.**
```
        1.06
 4.7)4.9 82
     4 7
       2 8
       0
       2 82
       2 82
```

**21.**
```
        89.
 0.06)5.34
      4 8
       54
       54
```

**22.**
```
        1.25
 0.48)0.60 00
      48
      12 0
       9 6
       2 40
       2 40
```

**23.**
```
      12.
 3.5)42.0
     35
      7 0
      7 0
```

**24.**
```
        5 03.
 4.73)2379.19
      2365
        14 1
        00 0
        14 19
        14 19
```

**25.**
```
       78.
 0.08)6.24
      5 6
       64
       64
```

**26.**
```
        1.6
 0.52)0.83 2
      52
      31 2
      31 2
```

**27.**
```
      8.3
 2.9)24.0 7
     23 2
       8 7
       8 7
```

**28.**
```
          225.
 0.618)139.050
       123 6
        15 45
        12 36
         3 090
         3 090
```

**29.**
```
      5 2.
 0.6)31.2
     30
      1 2
      1 2
```

**30.**
```
        8.5
 0.94)7.99 0
      7 52
        47 0
        47 0
```

**31.**
```
       3.18
 6.7)21.3 06
     20 1
      1 2 0
         6 7
         5 36
         5 36
```

**32.**
```
        5.19
 0.08)0.41 52
      40
       1 5
         8
        72
        72
```

**33.**
```
        0.016
 1.95)0.03 120
       1 95
       1 170
       1 170
```

**34.** $A = l \times w$
$112.5 = l \times 12.5$
divide by 12.5 to
find the length

```
         9.
 12.5)112.5
      112.5
```
9 feet

**35.**
```
         15 44.
 2.50)3860.00      1544 tickets sold
      250
      1360
      1250
       110 0
       100 0
        10 00
        10 00
```

**36.**
```
          4.347      4 basketballs
 14.95)$65.00000     can be bought
       59 80         with or $65.00.
        5 200
        4 485
          7150
          5980
         11700
         10465
```

**37.**
```
        53.7    gallons
 ×      12.4    miles per gallon
       21 4 8
      107 4
      537
      665.8 8  miles
```

60

## Area of a Circle (page 52)

1. $A \approx 3.14 \times 6 \times 6$
   $\approx 3.14 \times 36$
   $A \approx 113.04$ sq m

2. $A \approx 3.14 \times 1.7 \times 1.7$
   $\approx 3.14 \times 2.89$
   $A \approx 9.0746$ or 9.07 sq ft

3. $A \approx 3.14 \times 4 \times 4$
   $\approx 3.14 \times 16$
   $A \approx 50.24$ sq in.

4. $A \approx 3.14 \times 0.2 \times 0.2$
   $\approx 3.14 \times 0.04$
   $A \approx 0.1256$ or 0.13 sq cm

5. $A \approx 3.14 \times 1.3 \times 1.3$
   $\approx 3.14 \times 1.69$
   $A \approx 5.3066$ or 5.31 sq mm

6. $A \approx 3.14 \times 4.4 \times 4.4$
   $\approx 3.14 \times 19.36$
   $A \approx 60.7904$ or 60.79 sq m

7. $A \approx 3.14 \times 9.5 \times 9.5$
   $\approx 3.14 \times 90.25$
   $\approx 283.385$ or 283.39 sq m

8. $A \approx 3.14 \times 2.9 \times 2.9$
   $\approx 3.14 \times 8.41$
   $A \approx 26.4074$ or 26.41 sq mi

9. $A \approx 3.14 \times 1.3 \times 1.3$
   $\approx 3.14 \times 1.69$
   $A \approx 5.3066$ or 5.31 sq cm

10. $A \approx 3.14 \times 17 \times 17$
    $\approx 3.14 \times 289$
    $A \approx 907.40$ sq ft

11. $A \approx 3.14 \times 12 \times 12$
    $\approx 3.14 \times 144$
    $A \approx 452.16$ sq km

12. $A \approx 3.14 \times 66 \times 66$
    $\approx 3.14 \times 4356$
    $A \approx 13677.84$ sq yd

13. $A \approx 3.14 \times 18 \times 18$
    $\approx 3.14 \times 324$
    $A \approx 1017.36$ sq m

14. $A \approx 3.14 \times 3.3 \times 3.3$
    $\approx 3.14 \times 10.89$
    $A \approx 34.1946$ or 34.19 sq ft

15. $A \approx 3.14 \times 7.9 \times 7.9$
    $\approx 3.14 \times 62.41$
    $A \approx 195.9674$ or 195.97 sq dm

## Fractions to Decimals (pages 53–54)

1. $0.\overline{37}$
2. $0.\overline{4}$
3. $0.7\overline{3}$
4. $2.96\overline{12}$
5. $0.\overline{01}$
6. $0.92\overline{5}$

7. 2.166 or $2.1\overline{6}$
   0.6)1.3 000
   12
   1 0
   6
   40
   36
   40
   36

8. 63.55 or $63.\overline{5}$
   9)572.00
   54
   32
   27
   5 0
   4 5
   50
   45

9. 0.833 or $0.8\overline{3}$
   0.6)0.5 000
   48
   20
   18
   20
   18

10. 54.54 or $54.\overline{54}$
    11)600.00
    55
    50
    44
    6 0
    5 5
    50
    44

11. 20.833 or $20.8\overline{3}$
    24)500.000
    48
    20 0
    19 2
    80
    72
    80
    72

12. 13.3 or $13.\overline{3}$
    3)40.0
    3
    10
    9
    1 0
    9

13. 79.166 or $79.1\overline{6}$
    0.24)10.00 000
    16 8
    2 20
    2 16
    40
    24
    1 60
    1 44
    160
    144

14. 0.266 or $.2\overline{6}$
    0.3)0 080
    6
    20
    18
    20
    18

15. 0.25
    25)6.25
    5 0
    1 25
    1 25

16. 0.043
    6.1)0.2 623
    2 44
    183
    183

17. 6.
    0.57)3.42
    3 42

18. 0.66 or $0.\overline{6}$
    6)4.00
    3 6
    40
    36

19. 0.625
    8)5.000
    4 8
    20
    16
    40
    40

20. 0.733 or $0.7\overline{3}$
    15)11.000
    10 5
    50
    45
    50
    45

21. 3.5
    0.4)1.4 0
    1 2
    2 0
    2 0

22. 0.66 or $0.\overline{6}$
    12)8.00
    7 2
    80
    72

23. 5.4
    3.8)20.5 2
    19 0
    1 5 2
    1 5 2

24. 0.1545 or $0.15\overline{4}$
    11)1.7000
    11
    60
    55
    50
    44
    60
    55

25. 0.008008 or $.\overline{008}$
    999)8.000000
    7 992
    8000
    7992

26. 0.09
    0.72)0.06 48
    6 48

61

**27.** 
$$
\begin{array}{r}
0.8181 \text{ or } 0.\overline{81} \\
11\overline{)9.0000} \\
\underline{8\,8}\phantom{000} \\
20\phantom{00} \\
\underline{11}\phantom{00} \\
90\phantom{0} \\
\underline{88}\phantom{0} \\
20 \\
\underline{11} \\
\end{array}
$$

**28.** 
$$
\begin{array}{r}
2.7 \\
59\overline{)159.3} \\
\underline{118}\phantom{0} \\
41\,3 \\
\underline{41\,3} \\
\end{array}
$$

**29.** 
$$
\begin{array}{r}
0.833 \text{ or } 0.8\overline{3} \\
18\overline{)15.000} \\
\underline{14\,4}\phantom{00} \\
60\phantom{0} \\
\underline{54}\phantom{0} \\
60 \\
\underline{54} \\
\end{array}
$$

**30.** 
$$
\begin{array}{r}
0.77 \text{ or } 0.\overline{7} \\
9\overline{)7.00} \\
\underline{6\,3}\phantom{0} \\
70 \\
\underline{63} \\
\end{array}
$$

**31.** 
$$
\begin{array}{r}
0.875 \\
8\overline{)7.000} \\
\underline{6\,4}\phantom{00} \\
60\phantom{0} \\
\underline{56}\phantom{0} \\
40 \\
\underline{40} \\
\end{array}
$$

**32.** 
$$
\begin{array}{r}
0.277 \text{ or } 0.2\overline{7} \\
18\overline{)5.000} \\
\underline{3\,6}\phantom{00} \\
1\,40\phantom{0} \\
\underline{1\,26}\phantom{0} \\
140 \\
\underline{126} \\
\end{array}
$$

**33.** 
$$
\begin{array}{r}
0.4375 \\
16\overline{)7.0000} \\
\underline{6\,4}\phantom{000} \\
60\phantom{00} \\
\underline{48}\phantom{00} \\
120\phantom{0} \\
\underline{112}\phantom{0} \\
80 \\
\underline{80} \\
\end{array}
$$

**34.** 
$$
\begin{array}{r}
0.4166 \text{ or } 0.41\overline{6} \\
12\overline{)5.0000} \\
\underline{4\,8}\phantom{000} \\
20\phantom{00} \\
\underline{12}\phantom{00} \\
80\phantom{0} \\
\underline{72}\phantom{0} \\
80 \\
\underline{72} \\
\end{array}
$$

**35.** 
$$
\begin{array}{r}
0.52 \\
25\overline{)13.00} \\
\underline{12\,5}\phantom{0} \\
50 \\
\underline{50} \\
\end{array}
$$

**36.** 
$$
\begin{array}{r}
0.733 \text{ or } 0.7\overline{3} \\
15\overline{)11.000} \\
\underline{10\,5}\phantom{00} \\
50\phantom{0} \\
\underline{45}\phantom{0} \\
50 \\
\underline{45} \\
\end{array}
$$

**37.** 
$$
\begin{array}{r}
0.166 \text{ or } 0.1\overline{6} \\
6\overline{)1.000} \\
\underline{6}\phantom{000} \\
40\phantom{0} \\
\underline{36}\phantom{0} \\
40 \\
\underline{36} \\
\end{array}
$$

**38.** 
$$
\begin{array}{r}
0.6 \\
5\overline{)3.0} \\
\underline{3\,0} \\
\end{array}
$$

**39.** 
$$
\begin{array}{r}
0.66 \text{ or } 0.\overline{6} \\
3\overline{)2.00} \\
\underline{1\,8}\phantom{0} \\
20 \\
\underline{18} \\
\end{array}
$$

**40.** 
$$
\begin{array}{r}
0.5454 \text{ or } 0.\overline{54} \\
11\overline{)6.0000} \\
\underline{5\,5}\phantom{000} \\
50\phantom{00} \\
\underline{44}\phantom{00} \\
60\phantom{0} \\
\underline{55}\phantom{0} \\
50 \\
\underline{44} \\
\end{array}
$$

**41.** 
$$
\begin{array}{r}
0.55 \text{ or } 0.\overline{5} \\
9\overline{)5.00} \\
\underline{4\,5}\phantom{0} \\
50 \\
\underline{45} \\
\end{array}
$$

**42.** 
$$
\begin{array}{r}
0.9 \\
10\overline{)9.0} \\
\underline{9\,0} \\
\end{array}
$$

**43.** 
$$
\begin{array}{r}
0.25 \\
4\overline{)1.00} \\
\underline{8}\phantom{00} \\
20 \\
\underline{20} \\
\end{array}
$$

**44.** 
$$
\begin{array}{r}
0.633 \text{ or } 0.6\overline{3} \\
30\overline{)19.000} \\
\underline{18\,0}\phantom{00} \\
1\,00\phantom{0} \\
\underline{90}\phantom{0} \\
100 \\
\underline{90} \\
\end{array}
$$

**45.** 
$$
\begin{array}{r}
0.488 \text{ or } 0.4\overline{8} \\
45\overline{)22.000} \\
\underline{18\,0}\phantom{00} \\
4\,00\phantom{0} \\
\underline{3\,60}\phantom{0} \\
400 \\
\underline{360} \\
\end{array}
$$

**46.** 
$$
\begin{array}{r}
0.85 \\
20\overline{)17.00} \\
\underline{16\,0}\phantom{0} \\
1\,00 \\
\underline{1\,00} \\
\end{array}
$$

**47.** 
$$
\begin{array}{r}
0.525 \\
40\overline{)21.000} \\
\underline{20\,0}\phantom{00} \\
1\,00\phantom{0} \\
\underline{80}\phantom{0} \\
200 \\
\underline{200} \\
\end{array}
$$

**48.** 
$$
\begin{array}{r}
0.57142857 \text{ or } 0.\overline{571428} \\
7\overline{)4.00000000} \\
\underline{3\,5}\phantom{0000000} \\
50\phantom{000000} \\
\underline{49}\phantom{000000} \\
10\phantom{00000} \\
\underline{7}\phantom{00000} \\
30\phantom{0000} \\
\underline{28}\phantom{0000} \\
20\phantom{000} \\
\underline{14}\phantom{000} \\
60\phantom{00} \\
\underline{56}\phantom{00} \\
40\phantom{0} \\
\underline{35}\phantom{0} \\
50 \\
\underline{49} \\
\end{array}
$$

**49.** 
$$
\begin{array}{r}
0.5166 \text{ or } 0.51\overline{6} \\
60\overline{)31.0000} \\
\underline{30\,0}\phantom{000} \\
1\,00\phantom{00} \\
\underline{60}\phantom{00} \\
400\phantom{0} \\
\underline{360}\phantom{0} \\
400 \\
\underline{360} \\
\end{array}
$$

# PERCENTS

## Ratio and Proportion

A *ratio* is a comparison of two numbers. The ratio that compares 29 to 100 can be written in the following ways.

**29 out of 100**     **29 to 100**     **29:100**     $\frac{29}{100}$

*Change each ratio to a fraction in simplest form.*

**1.** 3 out of 4            **2.** 71:200            **3.** 4 to 30

**4.** 5:7                   **5.** 2:2               **6.** 3 out of 18

The ratios $\frac{44}{2}$ and $\frac{66}{3}$ are equivalent. A mathematical statement of two equivalent ratios is called a **proportion**.

$2 \times 66 = 132$

$44 \times 3 = 132$

**The cross products of a proportion are equal.**

*Use the following ratios to write a proportion.*

**7.** 1 out of 3 and 5 out of 15     **8.** 2 to 5 and 8 to 20     **9.** 3:4 and 9:12

**10.** 5 out of 7 and 10 out of 14   **11.** 30 to 55 and 6 to 11   **12.** 1:4 and 75:300

**13.** 6 out of 36 and 1 out of 6    **14.** 3 to 8 and 9 to 24    **15.** 12:5 and 36:15

*Do the ratios form a proportion? Write yes or no. Use cross products.*

**16.** $\frac{3}{8}$ and $\frac{6}{16}$     **17.** $\frac{3}{24}$ and $\frac{1}{8}$     **18.** $\frac{16}{7}$ and $\frac{7}{3}$     **19.** $\frac{2}{5}$ and $\frac{8}{20}$

**20.** $\frac{1}{3}$ and $\frac{6}{12}$     **21.** $\frac{3}{7}$ and $\frac{5}{9}$     **22.** $\frac{1}{3}$ and $\frac{3}{11}$     **23.** $\frac{5}{2}$ and $\frac{35}{14}$

**24.** $\frac{4}{10}$ and $\frac{12}{25}$    **25.** $\frac{6}{16}$ and $\frac{3}{8}$    **26.** $\frac{7}{9}$ and $\frac{6}{7}$     **27.** $\frac{3}{40}$ and $\frac{4.5}{60}$

**28.** $\frac{5}{15}$ and $\frac{5}{15}$    **29.** $\frac{1.3}{2}$ and $\frac{0.6}{0.1}$   **30.** $\frac{3}{10}$ and $\frac{7.5}{25}$   **31.** $\frac{5}{6}$ and $\frac{15}{18}$

Cross products can be used to find missing numbers in a proportion.

$$\frac{2}{3} = \frac{n}{12}$$ The cross products are $2 \times 12$ and $3 \times n$.

$2 \times 12 = 3 \times n$  The cross products of a proportion are equal.

$24 = 3 \times n$

$$\frac{24}{3} = \frac{3 \times n}{3}$$

$8 = n$ **The missing number is 8.**

*Find the missing number in each proportion.*

**32.** $\dfrac{5}{x} = \dfrac{10}{8}$     **33.** $\dfrac{a}{4} = \dfrac{9}{1}$     **34.** $\dfrac{3}{4} = \dfrac{n}{16}$     **35.** $\dfrac{3}{h} = \dfrac{9}{12}$

**36.** $\dfrac{1}{4} = \dfrac{k}{24}$     **37.** $\dfrac{3}{5} = \dfrac{12}{c}$     **38.** $\dfrac{t}{7} = \dfrac{1}{2}$     **39.** $\dfrac{x}{18} = \dfrac{8}{9}$

**40.** $\dfrac{4}{10} = \dfrac{22}{y}$     **41.** $\dfrac{26}{13} = \dfrac{2}{z}$     **42.** $\dfrac{6}{9} = \dfrac{b}{15}$     **43.** $\dfrac{7}{5} = \dfrac{d}{10}$

**44.** $\dfrac{1}{2} = \dfrac{m}{16}$     **45.** $\dfrac{1}{3} = \dfrac{f}{15}$     **46.** $\dfrac{2}{3} = \dfrac{m}{27}$     **47.** $\dfrac{1}{6} = \dfrac{13}{c}$

**48.** $\dfrac{5}{7} = \dfrac{40}{y}$     **49.** $\dfrac{f}{64} = \dfrac{1}{8}$     **50.** $\dfrac{5}{p} = \dfrac{20}{8}$     **51.** $\dfrac{8}{7} = \dfrac{32}{q}$

**52.** $\dfrac{18}{p} = \dfrac{3}{4}$     **53.** $\dfrac{3}{5} = \dfrac{r}{15}$     **54.** $\dfrac{2}{5} = \dfrac{24}{t}$     **55.** $\dfrac{1}{5} = \dfrac{w}{25}$

*For each of the following state a proportion which expresses the given sizes and costs.*

**56.** 16 ounces at $3
36 ounces at $x$ dollars

**57.** 2 packages at $8
9 packages at $y$ dollars

**58.** 6 boxes at $48
2 boxes at $g$ dollars

**59.** 1.5 gallons at $d$ dollars
3 gallons at $2.99

**60.** 48 ounces at $5.75
10 ounces at $x$ dollars

**61.** 5 liters at $4.25
$x$ liters at $13.00

**62.** 20 ounces at $3.30
75 ounces at $x$ dollars

**63.** 8 liters at $7.00
$y$ liters at $2.20

**64.** 470 bushels for 2.5 acres
$x$ bushels for 9 acres

*Solve each of the following.*

**65.** If 64 feet of rope weighs 20 pounds, how much will 80 feet of the same kind of rope weigh?

**66.** In 5 hours of driving Kate traveled 235 kilometers. If she travels at the same rate, in 11 hours what distance will she have driven?

**67.** Mr. Yang paid $200 for 600 square feet of roofing material. He still needs to purchase 240 square feet of material. How much can Mr. Yang expect to pay for this additional material?

**68.** The width of a photograph is 6 inches. Its length is 10 inches. The photograph is enlarged so that the width is 18 inches. What is the length of the enlarged photograph?

# Decimals to Percents

**Percent** means *per one hundred.* Percents are one way of expressing hundredths using the percent symbol (%).

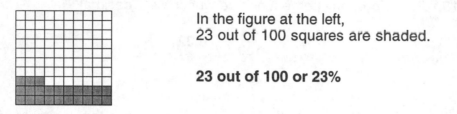

In the figure at the left,
23 out of 100 squares are shaded.

**23 out of 100 or 23%**

*Name the percent shown by each of the following.*

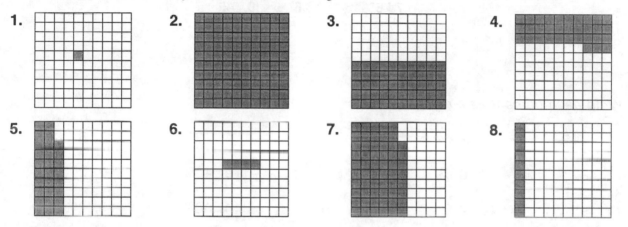

To change a decimal to a percent, move the decimal point two places to the right and annex the percent symbol (%).

$$0.35 \; y \;\; 0.35 \; y \;\; 35\%$$

*Change each decimal to a percent.*

| | | | | |
|---|---|---|---|---|
| **9.** 0.90 | **10.** 0.03 | **11.** 0.4 | **12.** 0.23 | **13.** 1.63 |
| **14.** 0.53 | **15.** 3.05 | **16.** 0.479 | **17.** 0.217 | **18.** 0.989 |
| **19.** 0.99 | **20.** 0.305 | **21.** 1.06 | **22.** 0.09 | **23.** 7.0 |
| **24.** 0.48 | **25.** 0.005 | **26.** 0.32 | **27.** 0.372 | **28.** 7.018 |
| **29.** 0.66 | **30.** 0.377 | **31.** 0.31 | **32.** 0.471 | **33.** 1.22 |
| **34.** 0.74 | **35.** 1.50 | **36.** 2.5 | **37.** 9.91 | **38.** 4.7 |
| **39.** 4.175 | **40.** 0.099 | **41.** 0.66 | **42.** 0.044 | **43.** 0.30 |

# Percents to Decimals

Because percent means *per one hundred*, any percent can be written as a ratio with a denominator of 100. Therefore, any percent can be written as a decimal.

$$32\% \quad y \quad \frac{32}{100} \quad y \quad 0.32$$

To write a percent as a decimal, remove the % symbol. Then, move the decimal point two places to the left.

$$74.5\% \quad y \quad 74.5 \quad y \quad 0.745$$

*Write each percent as a decimal.*

| | | | | |
|---|---|---|---|---|
| 1. 11% | 2. 28% | 3. 60.6% | 4. 79% | 5. 81% |
| 6. 87% | 7. 60% | 8. 6.5% | 9. 40% | 10. 300% |
| 11. 9% | 12. 2% | 13. 0.96% | 14. 30% | 15. 2% |
| 16. 707% | 17. 500% | 18. 400% | 19. 800% | 20. 830% |
| 21. 16.6% | 22. 90% | 23. 20.5% | 24. 62.7% | 25. 3.7% |
| 26. 4.1% | 27. 21% | 28. 9.9% | 29. 41.4% | 30. 8.7% |
| 31. 0.2% | 32. 2% | 33. 255.9% | 34. 123.4% | 35. 63.37% |
| 36. 4.13% | 37. 15.05% | 38. 0.57% | 39. 709.9% | 40. 2.59% |
| 41. 304.4% | 42. 20.55% | 43. 0.08% | 44. 36.71% | 45. 999.9% |

# Fractions to Percents, Percents to Fractions

Any percent can be written as a ratio with a denominator of 100.
A proportion can be used to change a fraction to a percent.

Change $\frac{3}{7}$ to a percent.

$$\frac{3}{7} = \frac{n}{100}$$ *Percents have a denominator of 100.*

$$3 \times 100 = 7n$$ *Cross products are equal.*

$$300 = 7n$$

$$\frac{300}{7} = \frac{7n}{7}$$ *Divide both sides by 7.*

$$42\frac{6}{7} = n$$

Since $n = 42\frac{6}{7}$ and $\frac{3}{7} = \frac{n}{100}$, then $\frac{3}{7} = \frac{42\frac{6}{7}}{100}$ or $42\frac{6}{7}\%$.

*Change each fraction to a percent.*

1. $\frac{1}{2}$      2. $\frac{3}{4}$      3. $\frac{3}{10}$      4. $\frac{1}{6}$      5. $\frac{9}{10}$

6. $\frac{4}{5}$      7. $\frac{1}{4}$      8. $\frac{2}{5}$      9. $\frac{1}{3}$      10. $\frac{3}{8}$

11. $\frac{7}{10}$      12. $\frac{1}{8}$      13. $\frac{?}{7}$      14. $\frac{4}{9}$      15. $\frac{5}{6}$

16. $\frac{3}{2}$      17. $\frac{4}{7}$      18. $\frac{7}{8}$      19. $\frac{2}{3}$      20. $\frac{6}{7}$

Fractions can also be changed to percents by dividing.
The resulting decimal is then changed to a percent.

The fraction $\frac{1}{6}$ is changed to a percent, rounded to the
nearest tenth in the example shown below.

$$\frac{1}{6} \rightarrow 6\overline{)1.0000}^{\,0.1666} \rightarrow 0.167, \text{ to nearest tenth} \rightarrow 16.7\%$$

*Change each fraction or mixed numeral to a percent rounded to the nearest tenth.*

21. $\frac{5}{6}$      22. $\frac{7}{8}$      23. $\frac{2}{3}$      24. $\frac{5}{9}$      25. $\frac{7}{16}$

26. $\frac{4}{7}$      27. $\frac{11}{12}$      28. $\frac{5}{12}$      29. $\frac{1}{8}$      30. $\frac{1}{3}$

31. $1\frac{9}{16}$      32. $1\frac{3}{8}$      33. $2\frac{3}{7}$      34. $2\frac{7}{12}$      35. $1\frac{2}{9}$

67

You can write percents as fractions by using the denominator of 100 and simplifying.

Write 55% as a fraction.　　**55%** → $\frac{55}{100}$　　　　　***Percent means per one hundred.***

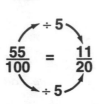

***Simplify.***

Write each percent as a fraction or mixed numeral in simplest form.

**36.** 7%　　　　**37.** 4%　　　　**38.** 6%　　　　**39.** 25%　　　　**40.** 50%

**41.** 1%　　　　**42.** 12%　　　**43.** 18%　　　**44.** 66%　　　　**45.** 65%

**46.** 60%　　　**47.** 38%　　　**48.** 250%　　**49.** 26%　　　　**50.** 10%

**51.** 85%　　　**52.** 15%　　　**53.** 22%　　　**54.** 58%　　　　**55.** 44%

**56.** 52%　　　**57.** 35.5%　　**58.** 27%　　　**59.** 120%　　　**60.** 360%

**61.** 80%　　　**62.** 5%　　　　**63.** 220%　　**64.** 78%　　　　**65.** 208%

# Fractions, Decimals, Percents

**Fractions to decimals.**

$$\frac{3}{5} \rightarrow 5\overline{)3} \rightarrow 0.6$$

**Fractions to percents.**

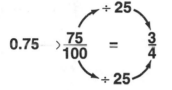

$$\frac{4}{9} \rightarrow \frac{4}{9} = \frac{n}{100}$$
$$4 \times 100 = 9 \times n$$
$$\frac{400}{9} = \frac{9n}{9}$$
$$44\frac{4}{9} = n \rightarrow 44\frac{4}{9}\%$$

**Decimals to fractions.**

$$0.75 \rightarrow \frac{75}{100} \overset{\div 25}{\underset{\div 25}{=}} \frac{3}{4}$$

**Decimals to percents.**

$$0.875 \ \text{y} \ 0.875 \ \text{y} \ 87.5\%$$

**Percents to decimals.**

$$6\% \ \text{y} \ 06\% \ \text{v} \ 0.06$$

**Percents to fractions.**

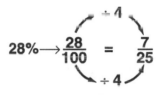

$$28\% \rightarrow \frac{28}{100} \overset{\div 4}{\underset{\div 4}{=}} \frac{7}{25}$$

*Change each fraction to a decimal.*

1. $\frac{4}{15}$  2. $\frac{7}{18}$  3. $\frac{2}{5}$  4. $\frac{3}{8}$  5. $\frac{11}{16}$

*Change each decimal to a fraction in simplest form.*

6. 0.64  7. 0.006  8. 5.18  9. 15.04  10. 0.202

*Change each decimal to a percent.*

11. 0.37  12. 0.50  13. 1.66  14. 0.07  15. 0.125

*Change each percent to a decimal.*

16. 45%  17. 8.1%  18. 20%  19. 810%  20. 26.4%

*Change each fraction to a percent.*

21. $\frac{1}{10}$  22. $\frac{5}{7}$  23. $\frac{1}{8}$  24. $\frac{4}{5}$  25. $\frac{5}{4}$

*Change each percent to a fraction in simplest form.*

26. 30%  27. 25%  28. 2%  29. 95%  30. 28%

# Finding the Percent of a Number

A percent problem can be solved using the following proportion.

$$\frac{\text{Percentage}}{\text{Base}} = \text{Rate or } \frac{P}{B} = \frac{r}{100}$$

The percentage ($P$) is the number which is compared to another number called the base ($B$). Rate is a percent. It follows that $r$ is always compared to 100.

**70% of 9 is what number?**

$$\frac{P}{B} = \frac{r}{100} \rightarrow \frac{P}{9} = \frac{70}{100}$$
$$P \times 100 = 9 \times 70$$
$$\frac{P \times 100}{100} = \frac{630}{100}$$
$$P = 6.3$$

70% of 9 is 6.3.

**What number is 15% of 500?**

$$\frac{P}{B} = \frac{r}{100} \rightarrow \frac{P}{500} = \frac{15}{100}$$
$$P \times 100 = 500 \times 15$$
$$\frac{P \times 100}{100} = \frac{7500}{100}$$
$$P = 75$$

75 is 15% of 500.

*Solve each of the following. Use the proportion given.*

1. _____ is 80% of 22.

$$\frac{P}{22} = \frac{80}{100}$$

2. _____ is 23% of 3100.

$$\frac{P}{3100} = \frac{23}{100}$$

3. 20% of 200 is _____.

$$\frac{P}{200} = \frac{20}{100}$$

*Name the percentage, base, and rate for each of the following.*

*Then solve using the proportion $\frac{r}{100} = \frac{P}{B}$.*

4. 10% of 120 is _____.

5. _____ is 19% of 300.

6. _____ is 85% of 60.

7. _____ is 10% of 55.

8. 8% of 125 is _____.

9. _____ is 40% of 35.

10. 6% of 80 is _____.

11. 8% of 70 is _____.

12. 11% of 20 is _____.

13. $66\frac{2}{3}$% of 225 is _____.

14. 50% of 21 is _____.

15. $37\frac{1}{2}$% of 64 is _____.

*Solve each problem.*

16. Sam's Sub Shop sold 225 submarine sandwiches Saturday. Italian submarine sandwiches accounted for 85% of their sales. How many Italian submarine sandwiches were sold?

17. Leah's soccer team has won 62.5% of its games. They have played 24 games. How many have they won?

18. In the 1976 Summer Olympics there were 613 medals awarded. The United States won about 15%. About how many is this?

19. Of 15,253 field goals attempted, Kareem Abdul-Jabbar made 54.9%. How many field goals did he make?

# Finding What Percentage One Number is of Another

A proportion can be used to find what percent one number is of another.

$$\frac{\text{Percentage}}{\text{Base}} = \text{Rate or } \frac{P}{B} = \frac{r}{100}$$

What percent of 20 is 17?

Here, 17 (the percentage) is being compared to 20 (the base).

$\frac{P}{B} = \frac{r}{100} \rightarrow \frac{17}{20} = \frac{r}{100}$ **Replace P by 17 and B by 20.**

$17 \times 100 = 20 \times r$ **Use the proportion rule.**

$\frac{1700}{20} = \frac{20 \times r}{20}$ **Divide both sides by 20.**

$85 = r$ **85% of 20 is 17.**

*Solve each of the following. Use the proportion given.*

1. 16 is _____% of 32.

$\frac{16}{32} = \frac{r}{100}$

2. 12.5 is _____% of 15.

$\frac{12.5}{15} = \frac{r}{100}$

3. _____% of 150 is 18.

$\frac{18}{150} = \frac{r}{100}$

*Name the percentage, base, and rate for each of the following.*
*Then solve using the proportion $\frac{P}{B} = \frac{r}{100}$.*

4. _____ % of 50 is 6.

5. 22.8 is _____% of 456.

6. 12 is _____% of 32.

7. 15 is _____% of 25.

8. 36 is _____% ot 400.

9. 6 is _____% of 120.

10. 16 is _____% of 80.

11. _____% of 150 is 108.

12. _____% of 1.7 is 0.85

13. 5.25 is _____% of 21.

14. 2 is _____% of 16.

15. 60 is _____% of 180.

*Solve each problem.*

16. Charles drove 675 miles of the 900-mile trip. What percent of the trip did he drive?

17. On a fishing trip, Denise's line had 25 bites. She caught 16 fish. On what percent of the bites did she catch fish?

18. On a fishing trip, Min had 20 bites and caught 8 fish. On what percent of the bites did she catch a fish?

19. The Zimmermans spent $1600 on vacation. $400 was spent on gasoline. What percent of the total was spent on gasoline.

# Finding A Number When a Percent is Given

Percent problems can be solved using $\frac{P}{B} = \frac{r}{100}$.

### 30% of what number is 189?

Find the number (the base) to which 189 (the percentage) is being compared. The rate is 30%.

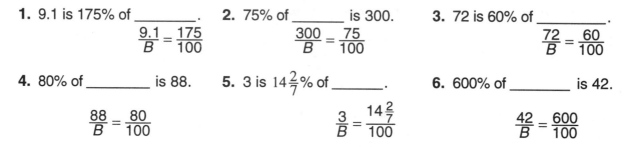

$$\frac{P}{B} = \frac{r}{100} \quad \rightarrow \quad \frac{189}{B} = \frac{30}{100} \qquad \text{Replace } P \text{ by 189 and } r \text{ by 30.}$$
$$189 \times 100 = B \times 30 \qquad \text{Use the proportion rule.}$$
$$\frac{18,900}{30} = \frac{B \times 30}{30} \qquad \text{Divide both sides by 30.}$$
$$630 = B \qquad \text{30\% of 630 is 189.}$$

*Solve each of the following. Use the proportion given.*

**1.** 9.1 is 175% of _____.
$$\frac{9.1}{B} = \frac{175}{100}$$

**2.** 75% of _____ is 300.
$$\frac{300}{B} = \frac{75}{100}$$

**3.** 72 is 60% of _____.
$$\frac{72}{B} = \frac{60}{100}$$

**4.** 80% of _____ is 88.
$$\frac{88}{B} = \frac{80}{100}$$

**5.** 3 is $14\frac{2}{7}$% of _____.
$$\frac{3}{B} = \frac{14\frac{2}{7}}{100}$$

**6.** 600% of _____ is 42.
$$\frac{42}{B} = \frac{600}{100}$$

*Name the percentage, base, and rate for each of the following.*
*Then solve using the proportion $\frac{P}{B} = \frac{r}{100}$.*

**7.** 56 is 70% of _____.

**8.** 15% of _____ is 3.

**9.** 28% of _____ is 5.6.

**10.** 8.5 is 50% of _____.

**11.** 6 is 1% of _____.

**12.** 20 is 100% of _____.

**13.** 30% of _____ is 60.

**14.** 18 is 30% of _____.

**15.** 125% of _____ is 35.

**16.** 75% of _____ is 72.

**17.** 99% of _____ is 0.891

**18.** 56 is 350% of _____.

*Solve each problem.*

**19.** Russ's Bait Shop can earn a profit if no more than 5% of the bait shipment arrive dead. Suppose 36 shrimp arrive dead. How many should be in Russ's shipment if it is accepted?

**20.** In 1978 Chris Evert won $120,411 in tournament play. This was 10% of the total winnings of the leading women players. What were the total winnings of leading women players?

# Percentage, Base, Rate

*Name the percentage (P), base (B), and rate (r) for each of the following.*

*Then, solve using the proportion* $\frac{P}{B} = \frac{r}{100}$.

1. 5% of 75 is _____.

2. 35% of 400 is _____.

3. _____% of 80 is 22.

4. _____% of 600 is 75.

5. _____ is 8% of 800.

6. 55% of _____ is 165.

7. 70% of 80 is _____.

8. _____ is 30% of 250.

9. 40% of _____ is 200.

10. 22 is _____% of 220.

11. 40 is _____% of 250.

12. 90 is 72% of _____.

13. _____% of 28 is 14.

14. 5% of 120 is _____.

15. 64% of _____ is 21.

16. _____% of 400 is 96.

17. $12\frac{1}{2}$% of _____ is 7.

18. 600% of _____ is 474.

19. 16 is 20% of _____.

20. $46\frac{2}{5}$% of 500 is _____.

21. _____% of 3600 is 9.

22. _____% of 120 is 30.

23. $87\frac{1}{2}$% of 160 is _____.

24. $22\frac{1}{2}$% of _____ is 31.5.

25. 5% of 456 is _____.

26. 794.5 is 350% of _____.

27. _____% of 80 is 25.

28. 1.1% of 2000 is _____.

29. 49 is _____% of 56.

30. 17.85 is _____% of 85.

31. $2\frac{1}{2}$% of 80 is _____.

32. 11.97 is 9% of _____.

33. _____ is 460% of 4.5

34. _____ is $4\frac{1}{4}$% of 600.

35. 3 is $14\frac{2}{7}$% of _____.

36. 48 is _____% of 120.

37. 30 is _____% of 80.

38. 43.5 is $7\frac{1}{2}$% of _____.

39. 6.38 is 145% of _____.

40. 511 is 7.3% of _____.

# Percent of Change (Gain/Loss)

A stereo system had a price change from $380 to $323.
Find the **percent of change** as follows.

Subtract to find the
amount of change.

$$\$380 - \$323 = \$57$$

To find the percent of
change, find what percent
$57 is of $380.

$$\frac{P}{B} = \frac{r}{100} \rightarrow \frac{57}{380} = \frac{r}{100}$$

$$5700 = 380 \times r$$

$$\frac{5700}{380} = \frac{380 \times r}{380}$$

$$15 = r$$

Compare $57 to $380.

The percent of change is a decrease of 15%.

*Find the amount of increase or decrease for each of the following.*

1. a $110 desk decreased to $99
2. a $40 clock radio decreased to $32
3. a $300 dishwasher increased to $315
4. an $80 CD player increased to $92
5. a $200 washer decreased to $160
6. a $250 dryer decreased to $200
7. a $60 radio decreased to $42
8. a $40 lamp increased to $44
9. a $150 guitar decreased to $135
10. a $20 calculator increased to $22

*Find the percent increase or decrease for each of the following.*

11. a $32.00 pair of sandals increased to $36.80
12. a $4 magazine reduced to $3
13. a $120 fax machine increased to $150
14. a $100 keyboard decreased to $72
15. a $6.00 book reduced to $4.20
16. a $250 microwave decreased to $200

*Solve each problem. Round to the nearest whole percent.*

17. A software package sold for $49. Now it is on sale for $43.12. What is the percent decrease?
18. The number of street lights was increased from 20 to 40. Find the percent of increase.
19. The price of a couch went from $1000 to $550. Find the percent decrease in price.
20. One year, a farm warehouse sold 120 tractors. The next year, they sold 150. What is the percent increase?
21. A microwave oven can cook a certain food in 7 minutes. A regular oven cooks the food in 35 minutes. What is the percent decrease in time?
22. A mountain bike is on sale for $429 from $550. What is the percent decrease in price?

# Discount

At the store, a television is on sale at a 15% **discount**. The original price was $738. Find the discount and the sale price.

To find 15% of $738, change 15% to a decimal and multiply by $738.

| $738 | *cost of TV* | | $738.00 | *cost* |
|---|---|---|---|---|
| × 0.15 | *(15% = 0.15)* | | − 110.70 | *discount* |
| $110.70 | *discount* | | $627.30 | *sale price* |

*A discount is an amount subtracted from the original price.*

The discount is $110.70. The sale price is $738 − $110.70 or $627.30.

*For each of the following, find the sale price given the original price and discount.*

1. original price, $45, $9 off

2. original price, $26.20, $3.54 off

3. original price, $18.99, $5.60 off

4. original price, $285.40, $29.80 off

5. original price, $29.75, $9.25 off

6. original price, $66.90, $8.50 off

*Find the discount on each item to the nearest cent.*

7. suit, $305
   discount rate, 20%

8. dress, $118.98
   discount rate, $33\frac{1}{3}$%

9. shoes, $51.92
   discount rate, $12\frac{1}{2}$%

10. jeans, $29.40
    discount rate, 5%

11. sunglasses, $209.90
    discount rate, 25%

12. shirt, $22.80
    discount rate, $10%

*Find the discount and sale price on each item to the nearest cent, given the original price and discount rate.*

13. chair, $100.95
    discount rate, $12\frac{1}{2}$%

14. couch, $316.00
    discount rate, 20%

15. book, $15.90
    discount rate, 12%

16. picture, $17.00
    discount rate, 11%

17. cabinet, $229.99
    discount rate, $33\frac{1}{3}$%

18. coffee table, $69.95
    discount rate, 10%

*Find the sale price for each of the following.*

19. price $110
    discount, $55

20. price, $3.72
    discount, $1.50

21. price, $75.78
    discount rate, $16\frac{2}{3}$%

22. price, $83.79
    discount rate, 35%

23. price, $19.80
    discount, $4.95

24. price, $655
    discount rate, 20%

25. price, $26
    discount rate, 40%

26. price, $79.89
    discount, $16.99

27. price, $31
    discount rate, 5%

28. price, $559.95
    discount, $17.76

29. price, $210
    discount rate, 10%

30. price, $407.50
    discount, $65.20

You can use a shortcut to find amounts such as a sale price.

Suppose an item costing $45 has a 12% discount. To find the sale price, you can solve the proportion at the right.
Explain why this method works.

**100% − 12% = 88% of original price.**

The sale price is $39.60.

$$\frac{P}{B} = \frac{r}{100}$$
$$\frac{P}{45} = \frac{88}{100}$$
$$P \times 100 = 45 \times 88$$
$$P \times 100 = 3960$$
$$\frac{P \times 100}{100} = \frac{3960}{100}$$
$$P = 39.60$$

*Use the shortcut method and $\frac{P}{B} = \frac{r}{100}$ to find the sale price.*

**31.** price, $1246
discount rate, 13%

**32.** price, $327
discount rate, 22%

**33.** price, $126
discount rate, 15%

**34.** price, $46
discount rate, 4%

**35.** price, $59.85
discount rate, 28%

**36.** price, $4560
discount rate, 18%

*Find the original price, discount, or discount rate, whichever is not given.*
*Round to the nearest cent.*

**37.** original price, $16
discount, $6.40

**38.** original price, $6.80
discount rate, 15%

**39.** original price, $19.95
discount, $7.98

**40.** original price, $12
discount rate, 35%

**41.** original price, $80
discount, $20

**42.** discount, $8.50
discount rate, 20%

**43.** discount, $29.40
discount rate, 60%

**44.** original price, $20
discount rate, 10%

**45.** discount, $9.75
discount rate, 25%

*Solve each problem.*

**46.** A washing machine costs $470. It is marked 20% off. Find the discount on the washing machine.

**47.** A stove usually costs $693.95. It is on sale at 25% off. Find the sale price.

**48.** A discount of $34.59 is offered on a $230.60 recorder. What is the discount rate?

**49.** A calculator is discounted $16\frac{2}{3}$%, or $8.25. What is the original price?

**50.** A radio originally costs $72.98. It is marked 10% off the original price. Find the discount on the radio. Then, find the sale price.

**51.** A microwave usually costs $579. It is on sale at a 15% discount. Find the discount.

# Profit (Markup)

Business firms keep a record of all their profits. **Profit** is the difference between the total income amount and the total costs of goods sold.

Suppose the total income amount is based on a 25% markup on the costs of goods sold. The **markup** is the dollar amount added to the cost to determine the selling price.

Suppose the cost of goods sold is $23,000; find the selling price.

Find the markup.

Add the markup to the cost to find the selling price.

To find the markup, multiply the % of markup by the cost of goods sold.

Then, add the markup to the cost.

$$\begin{array}{r} \$23{,}000 \quad \textit{cost of goods} \\ \times \quad 0.25 \quad \textit{(25\% = 0.25)} \\ \hline 1150\ 00 \\ 4600\ 0 \\ \hline \$5750.00 \quad \textit{markup} \end{array}$$

$$\begin{array}{r} \$23{,}000 \quad \textit{cost of goods} \\ + \quad 5750 \quad \textit{markup} \\ \hline \$28{,}750 \quad \textit{selling price of goods} \end{array}$$

The selling price is $28,750.

*Find the selling price for each of the following. The costs and percent of markup are given.*

1. $72,300, 25%

2. $38,000, 35%

3. 26,500, 20%

4. $30,950, 10%

5. $27,000, 15%

6. $40,000, 18%

Any percent problem can be solved using the proportion $\dfrac{P}{B} = \dfrac{r}{100}$.
Replace B by $23,000 and r by 25%.

$$\frac{P}{B} = \frac{r}{100}$$
$$\frac{P}{23{,}000} = \frac{25}{100}$$
$$P \times 100 = 23{,}000 \times 25$$
$$P \times 100 = 575{,}000$$
$$\frac{P \times 100}{100} = \frac{575{,}000}{100}$$
$$P = 5750$$

The selling price is
23,000 + 5750 or $28,750

*Use a proportion to solve each of the following.*

7. 100% of 115 is what number?

8. 25% of 60 is what number?

9. 70% of 90 is what number?

10. 75% of 600 is what number?

11. 40% of 35 is what number?

12. 42% of 500 is what number?

If the total costs of goods sold is $36,900 and the total income amount is $50,765, find the profit.

In order to find the profit, subtract the cost from the income.

$$\begin{array}{rl}
\$50,765 & \textit{income} \\
- \quad 36,900 & \textit{cost} \\
\hline
\$13,865 & \textit{profit}
\end{array}$$

The profit is $13,865.

*Solve. Round to the nearest cent.*

13. Denny's Hardware received a new shipment of tools. Each tool costs them $45.60. If Leckrone's use a 15% markup, what will be the selling price?

14. Leo sells his pottery on consignment at a store. If the pottery is sold, the store keeps a percentage of the selling price. Leo makes a bowl that sells for $27.50. The store keeps 35% of the selling price. Find how much the store keeps.

15. A clerk sells a $700 fax machine. The clerk earns 8% of the selling price. Find the earnings on the $700 sale.

16. A crate of strawberries at Peter's Fruit Stand costs the stand $56 per crate. If Peter's plans to sell a crate for $62.72, what is the percent of markup?

17. At Plash Department Store, the sales for July total $49,965. The expenses are $32,099. What are the profits for this month?

18. Farm and Ranch Supply wants a 30% markup on gates. A gate costs $139.95. Find the selling price.

# Depreciation

Most cars lose cash value as they get older. This is called **depreciation**.

The following percentages are average depreciation rates: 35% the first year, 17% the second year, 13% the third year, 11% the fourth year, and 8% the fifth year.

Suppose a new Chevette sells for $5468. How much does the car depreciate the first year?

To find the depreciation, multiply $5468 by 35%.

$$\begin{array}{r} \$5468 \\ \times\ \ \ 0.35 \\ \hline 273\ 40 \\ 1640\ 4 \\ \hline \$1913.80 \end{array}$$

*price of car*
*35%=0.35*

The Chevette loses $1913.80 in value in the first year.

Find the cash value of the car after 2 years.

$$\begin{array}{r} 35\% \quad \textit{first year} \\ +\ 17\% \quad \textit{second year} \\ \hline 52\% \quad \textit{depreciation} \end{array}$$

$100\% - 52\% = 48\%$

$$\begin{array}{r} \$5468 \\ \times\ \ \ 0.48 \\ \hline 437\ 44 \\ 2187\ 2 \\ \hline \$2624.64 \end{array}$$

*price of car*
*48% is 0.48*

The cash value after
2 years is $2624.64.

*Use the depreciation rates above. Answer each of the following questions.*

1. Robert bought a used car for $9360. How much cash value does it lose through depreciation in the fourth year?

2. Connie's car was priced at $12,632. After four years what is the cash value of her car?

3. A truck has a $21,780 price. Find the depreciation in the third year and find the cash value after 3 years.

4. Genya's first car cost $8109. How much cash value is lost through depreciation in the fifth year?

5. A new Geo costs $5872. How much of its cash value does it lose through depreciation in the third year?

6. A Ford Taurus costs $16,590. What is its cash value after 2 years?

7. Find the cash value after five years of an automobile which originally cost $11,723. What is the amount of depreciation over 5 years?

8. A Honda costs $6954. What is its cash value after each of the first 5 years?

# Simple Interest

Sergio's mother helped him get a loan of $75. He pays **interest** on the amount loaned. The interest rate is $9\frac{1}{2}\%$, or 9.5% per year. The time of the loan is 3 months, or $\frac{3}{12}$ years.

You can find the amount of interest on $75 for 3 months using the following formula.

$$interest = principal \times rate \times time$$
$$I = p \times r \times t$$

Principal ($p$) is the amount of the loan. The rate ($r$) is a percent, and time ($t$) is given in years.

$I = p \times r \times t$    *The principal is $75.*

$= 75 \times 0.095 \times \frac{3}{12}$    *The rate is 9.5% or 0.095.*

$= 75 \times 0.095 \times 0.25$    *The time, 3 months, is*

$= 1.78125$    *$\frac{3}{12}$ or $\frac{1}{4}$ or 0.25 years.*

To the nearest cent, the interest is $1.78.

In order to repay the loan, you must pay the principal plus interest. Therefore, the amount to be repaid is $75 + $1.78, or $76.78.

*Change months to years expressed as a decimal to the nearest hundredth.*

**1.** 6 months     **2.** 24 months     **3.** 12 months

**4.** 5 months     **5.** 2 months     **6.** 18 months

*Change each interest rate to a decimal.*

**7.** 17%    **8.** 11%    **9.** 5%    **10.** 2%    **11.** $14\frac{1}{4}\%$

**12.** $37\frac{1}{2}\%$    **13.** 11.33%    **14.** $6\frac{3}{4}\%$    **15.** $10\frac{3}{4}\%$    **16.** 41.75%

*Multiply to find the interest.*

**17.** $50 \times 0.10 \times 2$          **18.** $808 \times 0.07 \times 1$

**19.** $266 \times 0.12 \times 5$          **20.** $176 \times 0.08 \times 3$

**21.** $710 \times 0.11 \times 0.5$          **22.** $205 \times 0.40 \times 0.75$

*Find the interest to the nearest cent.*

**23.** $100 at 12% for 2 years          **24.** $200 at 15% for 6 years

**25.** $600 at $12\frac{1}{2}\%$ for 6 months      **26.** $2200 at $11\frac{1}{4}\%$ for 0.25 year

**27.** $50 for 15% for 0.5 year        **28.** $20 at 8% for 1 year

**29.** $1000 at 14% for 3 months     **30.** $800 at $14\frac{3}{4}\%$ for 2.5 years

# Taxes

Everyone who owns property must pay real estate taxes. Most people who work in arts and crafts must pay taxes on their homes, studios, and shops. The tax value of a house is a certain percent of its market value. Suppose a house is worth $70,000. Its tax value is 25% of the market value.

$$\frac{x}{70,000} = \frac{25}{100}$$
$$100x = 1,750,000$$
$$\frac{100x}{100} = \frac{1,750,000}{100}$$
$$x = 17,500$$

For tax purposes, the value of the house is $17,500. Now suppose the tax rate is 45.55 mills. In other words, for each $1000 of tax value, the owner must pay $45.55.

$$\frac{17,500}{1000} = 17.5 \qquad 17.5 \times 45.55 = 797.125 \qquad \text{The tax on the house is } \$797.13.$$

*The tax value of a house is 30% of its market value. Find the tax value for the given market values.*

1. $99,000
2. $175,000
3. $50,900
4. $78,800

5. $225,000
6. $460,000
7. $88,200
8. $97,200

*For the given tax values, find how many $1000 of tax value are taxed.*

9. $9500
10. $54,400
11. $9250
12. $22,440

13. $70,000
14. $107,000
15. $84,460
16. $99,970

*For the given $1000 of tax value, find the tax if the tax rate is 56.82 mills. Round to the nearest cent.*

17. 8.5
18. 7.7
19. 29.5
20. 39.4
21. 33.7

*Suppose the tax value of a house is 25% of the market value, and the tax rate is 45.55 mills. Find the tax on each market value to the nearest cent.*

22. $79,000
23. $59,900
24. $95,000
25. $45,000
26. $80,000

27. $72,500
28. $54,000
29. $30,000
30. $63,500
31. $34,900

*Solve each problem.*

32. The market value of a house is $68,500. The tax value is 35% of the market value. If the tax rate is 57.25 mills, what is the tax on the house?

33. The market value of a house is $54,000. The tax value is 30% of the market value. If the tax rate is 67.5 mills, what is the tax on the house?

81

When an item such as a walkie-talkie is sold, a **sales tax** may be charged. A 5% sales tax means a 5¢ tax is charged on each $1 of the purchase price.

Suppose the walkie-talkie costs $29.50. The sales tax rate is 5%.

Use the proportion $\frac{P}{B} = \frac{r}{100}$ to find the sales tax.

The base (*B*) is $29.50. The rate is 5%. The percentage (*P*) is the sales tax.

$$\frac{P}{B} = \frac{r}{100} \rightarrow \frac{P}{29.50} = \frac{5}{100}$$
$$P \times 100 = 147.50$$
$$\frac{P \times 100}{100} = \frac{147.50}{100}$$
$$P = 1.4750$$

To the nearest cent, the sales tax is $1.48.
The total price is $29.50 + $1.48 or $30.98.

*Find the sales tax to the nearest cent. Find the total price for each problem.*

**34.** price, $58

tax rate, $4\frac{1}{2}\%$

**35.** price, $265

tax rate, 5%

**36.** price, $18.50

tax rate, $6\frac{1}{4}\%$

**37.** price, $80

tax rate, 10%

**38.** price, $659.60

tax rate, $5\frac{1}{4}\%$

**39.** price, $6.59

tax rate, 9%

**40.** price, $124

tax rate, $8\frac{3}{4}\%$

**41.** price, $33

tax rate, 7%

**42.** price, $20

tax rate, 6%

*Complete the following with the word percentage, rate, or base.*

**43.** The sales tax rate is the _____ .

**44.** The purchase price is the _____ .    **45.** The amount of the sales tax is the _____ .

*For each of the following, find the price, sales tax, or tax rate, whichever is not given. Round to the nearest hundredth.*

**46.** price, $300

sales tax, $18

**47.** sales tax, $4.41

tax rate, $6\frac{1}{4}\%$

**48.** price, $212

sales tax, $13.25

**49.** sales tax, $1.50

tax rate, $7\frac{1}{2}\%$

**50.** price, $69.98

tax rate, $5\frac{3}{4}\%$

**51.** price, $26

sales tax, $2.02

**52.** price, $90
tax rate, 8%

**53.** sales tax, $9.81
tax rate, 4.5%

**54.** price, $221.40
tax rate, 4.6%

*Solve each problem.*

**55.** A coat costs $139.59. The tax rate is 5%. What is the sales tax on the coat?

**56.** A calculator costs $15.95. What is the sales tax to the nearest cent? The rate is 4.6%

**57.** A console stereo costs $266. The sales tax is $22.61. What is the sales tax rate?

**58.** The sales tax on a digital clock is $0.89. The tax rate is 5%. What is the price of the clock?

# ANSWERS and SOLUTIONS for PERCENTS

## Ratio and Proportion (pages 63–64)

1. $\frac{3}{4}$   2. $\frac{71}{200}$   3. $\frac{4}{30}=\frac{2}{15}$   4. $\frac{5}{7}$   5. $\frac{2}{2}=\frac{1}{1}$   6. $\frac{3}{18}=\frac{1}{6}$   7. $\frac{1}{3}=\frac{5}{15}$   8. $\frac{2}{5}=\frac{8}{20}$   9. $\frac{3}{4}=\frac{9}{12}$

10. $\frac{5}{7}=\frac{10}{14}$   11. $\frac{30}{55}=\frac{6}{11}$   12. $\frac{1}{4}=\frac{75}{300}$   13. $\frac{6}{36}=\frac{1}{6}$   14. $\frac{3}{8}=\frac{9}{24}$   15. $\frac{12}{5}=\frac{36}{15}$

16. $3\times16\overset{?}{=}6\times8$
    $48\overset{?}{=}48$, yes

17. $3\times8\overset{?}{=}1\times24$
    $24\overset{?}{=}24$, yes

18. $16\times3\overset{?}{=}7\times7$
    $48\overset{?}{=}49$, no

19. $2\times20\overset{?}{=}8\times5$
    $40\overset{?}{=}40$, yes

20. $1\times12\overset{?}{=}6\times3$
    $12\overset{?}{=}18$, no

21. $3\times9\overset{?}{=}5\times7$
    $27\overset{?}{=}35$, no

22. $1\times11\overset{?}{=}3\times3$
    $11\overset{?}{=}9$, no

23. $5\times14\overset{?}{=}35\times2$
    $70\overset{?}{=}70$, yes

24. $4\times25\overset{?}{=}12\times10$
    $100\overset{?}{=}120$, no

25. $6\times8\overset{?}{=}3\times16$
    $48\overset{?}{=}48$, yes

26. $7\times7\overset{?}{=}6\times9$
    $49\overset{?}{=}54$, no

27. $3\times60\overset{?}{=}4.5\times10$
    $180\overset{?}{=}180$, yes

28. $5\times15\overset{?}{=}5\times15$
    $75\overset{?}{=}75$, yes

29. $1.3\times0.1\overset{?}{=}0.6\times2$
    $0.13\overset{?}{=}1.2$, no

30. $3\times25\overset{?}{=}7.5\times10$
    $75\overset{?}{=}75$, yes

31. $5\times18\overset{?}{=}6\times15$
    $90\overset{?}{=}90$, yes

32. $5\times8=10\times x$
    $\frac{40}{10}=\frac{10\times x}{10}$
    $4=x$

33. $a\times1=9\times4$
    $\frac{a\times1}{1}=\frac{36}{1}$
    $a=36$

34. $3\times16=4\times n$
    $\frac{48}{4}=\frac{4\times n}{4}$
    $12=n$

35. $3\times12=9\times h$
    $\frac{36}{9}=\frac{9\times h}{9}$
    $4=h$

36. $1\times24=4\times k$
    $\frac{24}{4}=\frac{4\times k}{4}$
    $6=k$

37. $3\times c=5\times12$
    $\frac{3\times c}{3}=\frac{60}{3}$
    $c=20$

38. $2\times t=1\times7$
    $\frac{2\times t}{2}=\frac{7}{2}$
    $t=3\frac{1}{2}$

39. $8\times18=9\times x$
    $\frac{144}{9}=\frac{9\times x}{9}$
    $16=x$

40. $1\times y=22\times10$
    $\frac{4\times y}{4}=\frac{220}{4}$
    $y=55$

41. $26\times z=2\times13$
    $\frac{26\times z}{26}=\frac{26}{26}$
    $z=1$

42. $6\times15=9\times b$
    $\frac{90}{9}=\frac{9\times b}{9}$
    $10=b$

43. $7\times10=5\times d$
    $\frac{70}{5}=\frac{5\times d}{5}$
    $14=d$

44. $1\times16=2\times m$
    $\frac{16}{2}=\frac{2\times m}{2}$
    $8=m$

45. $1\times15=3\times f$
    $\frac{15}{3}=\frac{3\times f}{3}$
    $5=f$

46. $2\times27=3\times m$
    $\frac{54}{3}=\frac{3\times m}{3}$
    $10=m$

47. $1\times c=6\times13$
    $\frac{1\times c}{1}=\frac{78}{1}$
    $c=78$

48. $5\times y=7\times40$
    $\frac{5\times y}{5}=\frac{280}{5}$
    $y=56$

49. $1\times64=8\times t$
    $\frac{64}{8}=\frac{8\times f}{8}$
    $8=f$

50. $5\times8=20\times p$
    $\frac{40}{20}=\frac{20\times p}{20}$
    $2=p$

51. $8\times q=7\times32$
    $\frac{8\times q}{8}=\frac{224}{8}$
    $q=28$

52. $3\times p=18\times4$
    $\frac{3\times p}{3}=\frac{72}{3}$
    $p=24$

53. $3\times15=5\times r$
    $\frac{45}{5}=\frac{5\times r}{5}$
    $9=r$

54. $2\times t=5\times24$
    $\frac{2\times t}{2}=\frac{120}{2}$
    $t=60$

55. $1\times25=5\times w$
    $\frac{25}{5}=\frac{5\times w}{5}$
    $5=w$

56. $\frac{16}{3}=\frac{36}{x}$   57. $\frac{2}{8}=\frac{9}{y}$   58. $\frac{6}{48}=\frac{2}{g}$   59. $\frac{1.5}{d}=\frac{3}{2.99}$   60. $\frac{48}{5.75}=\frac{10}{x}$

61. $\frac{5}{4.25}=\frac{x}{13}$   62. $\frac{20}{3.3}=\frac{75}{x}$   63. $\frac{8}{7}=\frac{y}{2.20}$   64. $\frac{470}{2.5}=\frac{x}{9}$

65. $\frac{64}{20}=\frac{80}{w}$
    $64\times w=80\times20$
    $\frac{64\times w}{64}=\frac{1600}{64}$
    $w=25$ pounds

66. $\frac{5}{235}=\frac{11}{d}$
    $5\times d=11\times235$
    $\frac{5\times d}{5}=\frac{2585}{5}$
    $d=517$ km

67. $\frac{200}{600}=\frac{c}{240}$
    $200\times240=600\times c$
    $\frac{48,000}{600}=\frac{600\times c}{600}$
    $\$80=c$

68. $\frac{6}{10}=\frac{18}{\ell}$
    $6\times\ell=18\times10$
    $\frac{6\times\ell}{6}=\frac{180}{6}$
    $\ell=30$ in.

## Decimals to Percents (page 65)

1. 1%   2. 100%   3. 50%   4. 33%   5. 28%   6. 4%   7. 58%   8. 10%   9. 90%   10. 3%   11. 40%
12. 23%   13. 163%   14. 53%   15. 305%   16. 47.9%   17. 21.7%   18. 98.9%   19. 99%   20. 30.5%
21. 106%   22. 9%   23. 700%   24. 48%   25. 0.5%   26. 32%   27. 37.2%   28. 701.8%   29. 66%
30. 37.7%   31. 31%   32. 47.1%   33. 122%   34. 74%   35. 150%   36. 250%   37. 991%   38. 470%
39. 417.5%   40. 9.9%   41. 66%   42. 4.4%   43. 30%

## Percents to Decimals (page 66)

| | | | | | | | |
|---|---|---|---|---|---|---|---|
| **1.** 0.11 | **2.** 0.28 | **3.** 0.606 | **4.** 0.79 | **5.** 0.81 | **6.** 0.87 | **7.** 0.6 | **8.** 0.065 |
| **9.** 0.4 | **10.** 3.00 | **11.** 0.09 | **12.** 0.02 | **13.** 0.0096 | **14.** 0.3 | **15.** 0.02 | **16.** 7.07 |
| **17.** 5.00 | **18.** 4.00 | **19.** 8.00 | **20.** 8.30 | **21.** 0.166 | **22.** 0.90 | **23.** 0.205 | **24.** 0.627 |
| **25.** 0.037 | **26.** 0.041 | **27.** 0.21 | **28.** 0.099 | **29.** 0.414 | **30.** 0.087 | **31.** 0.002 | **32.** 0.02 |
| **33.** 2.559 | **34.** 1.234 | **35.** 0.6337 | **36.** 0.0413 | **37.** 0.1505 | **38.** 0.0057 | **39.** 7.099 | **40.** 0.0259 |
| **41.** 3.044 | **42.** 0.2055 | **43.** 0.0008 | **44.** 0.3671 | **45.** 9.999 | | | |

## Fractions to Percents, Percents to Fractions (pages 67–68)

**1.**
$$\frac{1}{2} = \frac{n}{100}$$
$$1 \times 100 = 2 \times n$$
$$\frac{100}{2} = \frac{2 \times n}{2}$$
$$50 = n, 50\%$$

**2.**
$$\frac{3}{4} = \frac{n}{100}$$
$$3 \times 100 = 4 \times n$$
$$\frac{300}{4} = \frac{4 \times n}{4}$$
$$75 = n, 75\%$$

**3.**
$$\frac{3}{10} = \frac{n}{100}$$
$$3 \times 100 = 10 \times n$$
$$\frac{300}{10} = \frac{10 \times n}{10}$$
$$30 = n, 30\%$$

**4.**
$$\frac{1}{5} = \frac{n}{100}$$
$$1 \times 100 = 5 \times n$$
$$\frac{100}{5} = \frac{5 \times n}{5}$$
$$20 = n, 20\%$$

**5.**
$$\frac{9}{10} = \frac{n}{100}$$
$$9 \times 100 = 10 \times n$$
$$\frac{900}{10} = \frac{10 \times n}{10}$$
$$90 = n, 90\%$$

**6.**
$$\frac{4}{5} = \frac{n}{100}$$
$$4 \times 100 = 5 \times n$$
$$\frac{400}{5} = \frac{5 \times n}{5}$$
$$80 = n, 80\%$$

**7.**
$$\frac{1}{4} = \frac{n}{100}$$
$$100 \times 1 = 4 \times n$$
$$\frac{100}{4} = \frac{4 \times n}{4}$$
$$25 = n, 25\%$$

**8.**
$$\frac{2}{5} = \frac{n}{100}$$
$$2 \times 100 = 5 \times n$$
$$\frac{200}{5} = \frac{5 \times n}{5}$$
$$40 = n, 40\%$$

**9.**
$$\frac{1}{3} = \frac{n}{100}$$
$$1 \times 100 = 3 \times n$$
$$\frac{100}{3} = \frac{3 \times n}{3}$$
$$33.\overline{3} \text{ or } 33\frac{1}{3} = n, 33\frac{1}{3}\%$$

**10.**
$$\frac{3}{8} = \frac{n}{100}$$
$$3 \times 100 = 8 \times n$$
$$\frac{300}{8} = \frac{8 \times n}{8}$$
$$37.5 \text{ or } 37\frac{1}{2} = n, 37.5\%$$

**11.**
$$\frac{7}{10} = \frac{n}{100}$$
$$7 \times 100 = 10 \times n$$
$$\frac{700}{10} = \frac{10 \times n}{10}$$
$$70 = n, 70\%$$

**12.**
$$\frac{1}{8} = \frac{n}{100}$$
$$1 \times 100 = 8 \times n$$
$$\frac{100}{8} = \frac{8 \times n}{8}$$
$$12.5 \text{ or } 12\frac{1}{2} = n, 12.5\%$$

**13.**
$$\frac{2}{7} = \frac{n}{100}$$
$$2 \times 100 = 7 \times n$$
$$\frac{200}{7} = \frac{7 \times n}{7}$$
$$28\frac{4}{7} = n, 28\frac{4}{7}\%$$

**14.**
$$\frac{4}{9} = \frac{n}{100}$$
$$4 \times 100 = 9 \times n$$
$$\frac{400}{9} = \frac{9 \times n}{9}$$
$$44.\overline{4} \text{ or } 44\frac{4}{9} = n, 44.\overline{4}\%$$

**15.**
$$\frac{5}{6} = \frac{n}{100}$$
$$5 \times 100 = 6 \times n$$
$$\frac{500}{6} = \frac{6 \times n}{6}$$
$$83.\overline{3} \text{ or } 83\frac{1}{3} = n, 83\frac{1}{3}\%$$

**16.**
$$\frac{3}{2} = \frac{n}{100}$$
$$3 \times 100 = 2 \times n$$
$$\frac{300}{2} = \frac{2 \times n}{2}$$
$$150 = n, 150\%$$

**17.**
$$\frac{4}{7} = \frac{n}{100}$$
$$4 \times 100 = 7 \times n$$
$$\frac{400}{7} = \frac{7 \times n}{7}$$
$$57\frac{1}{7} = n, 57\frac{1}{7}\%$$

**18.**
$$\frac{7}{8} = \frac{n}{100}$$
$$7 \times 100 = 8 \times n$$
$$\frac{700}{8} = \frac{8 \times n}{8}$$
$$87.5 \text{ or } 87\frac{1}{2} = n, 87.5\%$$

**19.**
$$\frac{2}{3} = \frac{n}{100}$$
$$2 \times 100 = 3 \times n$$
$$\frac{200}{3} = \frac{3 \times n}{3}$$
$$66.\overline{6} \text{ or } 66\frac{2}{3} = n, 66\frac{2}{3}\%$$

**20.**
$$\frac{6}{7} = \frac{n}{100}$$
$$6 \times 100 = 7 \times n$$
$$\frac{600}{7} = \frac{7 \times n}{7}$$
$$85\frac{5}{7} = n, 85\frac{5}{7}\%$$

**21.**
```
      0.8333  y 83.3%
  6)5.0000
    4 8
    ‾‾‾
     20
     18
     ‾‾
      20
      18
      ‾‾
       20
       18
```

**22.**
```
      0.875  y 87.5%
  8)7.000
    6 4
    ‾‾‾
     60
     56
     ‾‾
      40
      40
```

**23.**
```
      0.6666  y 66.7%
  3)2.0000
    1 8
    ‾‾‾
     20
     18
     ‾‾
      20
      18
      ‾‾
       20
       18
```

**24.**
```
      0.5555  y 55.6%
  9)5.0000
    4 5
    ‾‾‾
     50
     45
     ‾‾
      50
      45
      ‾‾
       50
       45
```

**25.**
```
      0.4375  y 43.8%
  16)7.0000
     6 4
     ‾‾‾
      60
      48
      ‾‾
      120
      112
      ‾‾‾
       80
       80
```

**26.**
```
      0.5714  y 57.1%
  7)4.0000
    3 5
    ‾‾‾
     50
     49
     ‾‾
      10
       7
      ‾‾
       30
       28
```

84

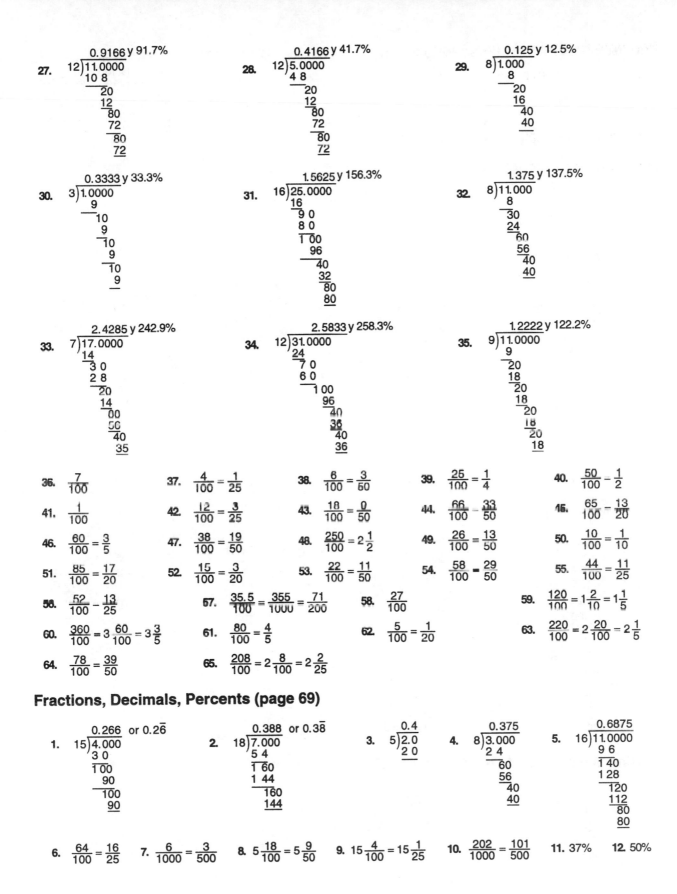

27. $\dfrac{0.9166}{12)11.0000}$ y 91.7%

$\dfrac{10\ 8}{\phantom{0}20}$
$\dfrac{12}{\phantom{0}80}$
$\dfrac{72}{\phantom{0}80}$
$\dfrac{72}{\phantom{0}}$

28. $\dfrac{0.4166}{12)5.0000}$ y 41.7%

$\dfrac{4\ 8}{\phantom{0}20}$
$\dfrac{12}{\phantom{0}80}$
$\dfrac{72}{\phantom{0}80}$
$\dfrac{72}{\phantom{0}}$

29. $\dfrac{0.125}{8)1.000}$ y 12.5%

$\dfrac{8}{20}$
$\dfrac{16}{40}$
$\dfrac{40}{\phantom{0}}$

30. $\dfrac{0.3333}{3)1.0000}$ y 33.3%

$\dfrac{9}{10}$
$\dfrac{9}{10}$
$\dfrac{9}{10}$
$\dfrac{9}{\phantom{0}}$

31. $\dfrac{1.5625}{16)25.0000}$ y 156.3%

$\dfrac{16}{9\ 0}$
$\dfrac{8\ 0}{1\ 00}$
$\dfrac{96}{40}$
$\dfrac{32}{80}$
$\dfrac{80}{\phantom{0}}$

32. $\dfrac{1.375}{8)11.000}$ y 137.5%

$\dfrac{8}{30}$
$\dfrac{24}{60}$
$\dfrac{56}{40}$
$\dfrac{40}{\phantom{0}}$

33. $\dfrac{2.4285}{7)17.0000}$ y 242.9%

$\dfrac{14}{3\ 0}$
$\dfrac{2\ 8}{20}$
$\dfrac{14}{00}$
$\dfrac{50}{40}$
$\dfrac{35}{\phantom{0}}$

34. $\dfrac{2.5833}{12)31.0000}$ y 258.3%

$\dfrac{24}{7\ 0}$
$\dfrac{6\ 0}{1\ 00}$
$\dfrac{96}{40}$
$\dfrac{36}{40}$
$\dfrac{36}{\phantom{0}}$

35. $\dfrac{1.2222}{9)11.0000}$ y 122.2%

$\dfrac{9}{20}$
$\dfrac{18}{20}$
$\dfrac{18}{20}$
$\dfrac{18}{20}$
$\dfrac{18}{\phantom{0}}$

36. $\dfrac{7}{100}$

37. $\dfrac{4}{100} = \dfrac{1}{25}$

38. $\dfrac{6}{100} = \dfrac{3}{50}$

39. $\dfrac{25}{100} = \dfrac{1}{4}$

40. $\dfrac{50}{100} = \dfrac{1}{2}$

41. $\dfrac{1}{100}$

42. $\dfrac{12}{100} = \dfrac{3}{25}$

43. $\dfrac{18}{100} = \dfrac{9}{50}$

44. $\dfrac{66}{100} = \dfrac{33}{50}$

45. $\dfrac{65}{100} = \dfrac{13}{20}$

46. $\dfrac{60}{100} = \dfrac{3}{5}$

47. $\dfrac{38}{100} = \dfrac{19}{50}$

48. $\dfrac{250}{100} = 2\dfrac{1}{2}$

49. $\dfrac{26}{100} = \dfrac{13}{50}$

50. $\dfrac{10}{100} = \dfrac{1}{10}$

51. $\dfrac{85}{100} = \dfrac{17}{20}$

52. $\dfrac{15}{100} = \dfrac{3}{20}$

53. $\dfrac{22}{100} = \dfrac{11}{50}$

54. $\dfrac{58}{100} = \dfrac{29}{50}$

55. $\dfrac{44}{100} = \dfrac{11}{25}$

56. $\dfrac{52}{100} = \dfrac{13}{25}$

57. $\dfrac{35.5}{100} = \dfrac{355}{1000} = \dfrac{71}{200}$

58. $\dfrac{27}{100}$

59. $\dfrac{120}{100} = 1\dfrac{2}{10} = 1\dfrac{1}{5}$

60. $\dfrac{360}{100} = 3\dfrac{60}{100} = 3\dfrac{3}{5}$

61. $\dfrac{80}{100} = \dfrac{4}{5}$

62. $\dfrac{5}{100} = \dfrac{1}{20}$

63. $\dfrac{220}{100} = 2\dfrac{20}{100} = 2\dfrac{1}{5}$

64. $\dfrac{78}{100} = \dfrac{39}{50}$

65. $\dfrac{208}{100} = 2\dfrac{8}{100} = 2\dfrac{2}{25}$

## Fractions, Decimals, Percents (page 69)

1. $\dfrac{0.266}{15)4.000}$ or $0.2\overline{6}$

$\dfrac{3\ 0}{100}$
$\dfrac{90}{100}$
$\dfrac{90}{\phantom{0}}$

2. $\dfrac{0.388}{18)7.000}$ or $0.3\overline{8}$

$\dfrac{5\ 4}{1\ 60}$
$\dfrac{1\ 44}{160}$
$\dfrac{144}{\phantom{0}}$

3. $\dfrac{0.4}{5)2.0}$

$\dfrac{2\ 0}{\phantom{0}}$

4. $\dfrac{0.375}{8)3.000}$

$\dfrac{2\ 4}{60}$
$\dfrac{56}{40}$
$\dfrac{40}{\phantom{0}}$

5. $\dfrac{0.6875}{16)11.0000}$

$\dfrac{9\ 6}{1\ 40}$
$\dfrac{1\ 28}{120}$
$\dfrac{112}{80}$
$\dfrac{80}{\phantom{0}}$

6. $\dfrac{64}{100} = \dfrac{16}{25}$

7. $\dfrac{6}{1000} = \dfrac{3}{500}$

8. $5\dfrac{18}{100} = 5\dfrac{9}{50}$

9. $15\dfrac{4}{100} = 15\dfrac{1}{25}$

10. $\dfrac{202}{1000} = \dfrac{101}{500}$

11. 37%

12. 50%

85

**13.** 166%  **14.** 7%  **15.** 12.5%  **16.** 0.45  **17.** 0.081  **18.** 0.2  **19.** 8.1  **20.** 0.264  **21.**
$$\frac{1}{10} = \frac{n}{100}$$
$$1 \times 100 = 10 \times n$$
$$\frac{100}{10} = \frac{10 \times n}{10}$$
$$10 = n,\ 10\%$$

**22.**
$$\frac{5}{7} = \frac{n}{100}$$
$$5 \times 100 = 7 \times n$$
$$\frac{500}{7} = \frac{7 \times n}{7}$$
$$71.4 \text{ or } 71\tfrac{3}{7} = n,\ 71\tfrac{3}{7}\%$$

**23.**
$$\frac{1}{8} = \frac{n}{100}$$
$$1 \times 100 = 8 \times n$$
$$\frac{100}{8} = \frac{8 \times n}{8}$$
$$12.5 \text{ or } 12\tfrac{1}{2} = n,\ 12.5\%$$

**24.**
$$\frac{4}{5} = \frac{n}{100}$$
$$4 \times 100 = 5 \times n$$
$$\frac{400}{5} = \frac{5 \times n}{5}$$
$$80 = n,\ 80\%$$

**25.**
$$\frac{5}{4} = \frac{n}{100}$$
$$5 \times 100 = 4 \times n$$
$$\frac{500}{4} = \frac{4 \times n}{4}$$
$$125 = n,\ 125\%$$

**26.** $\dfrac{30}{100} = \dfrac{3}{10}$   **27.** $\dfrac{25}{100} = \dfrac{1}{4}$   **28.** $\dfrac{2}{100} = \dfrac{1}{50}$   **29.** $\dfrac{95}{100} = \dfrac{19}{20}$   **30.** $\dfrac{28}{100} = \dfrac{7}{25}$

## Finding the Percent of a Number (page 70)

**1.**
$$\frac{P}{22} = \frac{80}{100}$$
$$100 \times P = 22 \times 80$$
$$\frac{100 \times P}{100} = \frac{1760}{100}$$
$$P = 17.6$$

**2.**
$$\frac{P}{3100} = \frac{23}{100}$$
$$100 \times P = 23 \times 3100$$
$$\frac{100 \times P}{100} = \frac{71,300}{100}$$
$$P = 713$$

**3.**
$$\frac{P}{200} = \frac{20}{100}$$
$$100 \times P = 20 \times 200$$
$$\frac{100 \times P}{100} = \frac{4000}{100}$$
$$P = 40$$

**4.** $r = 10,\ B = 120$
$$\frac{10}{100} = \frac{P}{120}$$
$$10 \times 120 = 100 \times P$$
$$\frac{1200}{100} = \frac{100 \times P}{100}$$
$$12 = P$$

**5.** $r = 19,\ B = 300$
$$\frac{19}{100} = \frac{P}{300}$$
$$19 \times 300 = 100 \times P$$
$$\frac{5700}{100} = \frac{100 \times P}{100}$$
$$57 = P$$

**6.** $r = 85,\ B = 60$
$$\frac{85}{100} = \frac{P}{60}$$
$$85 \times 60 = 100 \times P$$
$$\frac{5100}{100} = \frac{100 \times P}{100}$$
$$51 = P$$

**7.** $r = 10,\ B = 55$
$$\frac{10}{100} = \frac{P}{55}$$
$$10 \times 55 = 100 \times P$$
$$\frac{550}{100} = \frac{100 \times P}{100}$$
$$5.5 = P$$

**8.** $r = 8,\ B = 125$
$$\frac{8}{100} = \frac{P}{125}$$
$$8 \times 125 = 100 \times P$$
$$\frac{1000}{100} = \frac{100 \times P}{100}$$
$$10 = P$$

**9.** $r = 40,\ B = 35$
$$\frac{40}{100} = \frac{P}{35}$$
$$40 \times 35 = 100 \times P$$
$$\frac{1400}{100} = \frac{100 \times P}{100}$$
$$14 = P$$

**10.** $r = 6,\ B = 80$
$$\frac{6}{100} = \frac{P}{80}$$
$$6 \times 80 = 100 \times P$$
$$\frac{480}{100} = \frac{100 \times P}{100}$$
$$4.8 = P$$

**11.** $r = 8,\ B = 70$
$$\frac{8}{100} = \frac{P}{70}$$
$$8 \times 70 = 100 \times P$$
$$\frac{560}{100} = \frac{100 \times P}{100}$$
$$5.6 = P$$

**12.** $r = 11,\ B = 20$
$$\frac{11}{100} = \frac{P}{20}$$
$$20 \times 11 = 100 \times P$$
$$\frac{220}{100} = \frac{100 \times P}{100}$$
$$2.2 = P$$

**13.** $r = 66\tfrac{2}{3},\ B = 225$
$$\frac{66\tfrac{2}{3}}{100} = \frac{P}{225}$$
$$\frac{200}{3} \times 225 = 100 \times P$$
$$\frac{15,000}{100} = \frac{100 \times P}{100}$$
$$150 = P$$

**14.** $r = 50,\ B = 21$
$$\frac{50}{100} = \frac{P}{21}$$
$$50 \times 21 = 100 \times P$$
$$\frac{1050}{100} = \frac{100 \times P}{100}$$
$$10.5 = P$$

**15.** $r = 37\tfrac{1}{2},\ B = 64$
$$\frac{37\tfrac{1}{2}}{100} = \frac{P}{64}$$
$$\frac{75}{2} \times 64 = 100 \times P$$
$$\frac{2400}{100} = \frac{100 \times P}{100}$$
$$24 = P$$

**16.**
$$\frac{85}{100} = \frac{P}{225}$$
$$85 \times 225 = 100 \times P$$
$$\frac{19,125}{100} = \frac{100 \times P}{100}$$
$$191.25 = P$$
or about 191 Italian submarine sandwiches

**17.**
$$\frac{62.5}{100} = \frac{P}{24}$$
$$62.5 \times 24 = 100 \times P$$
$$\frac{1500}{100} = \frac{100 \times P}{100}$$
$$15 = P$$
15 games won

**18.**
$$\frac{15}{100} = \frac{P}{613}$$
$$15 \times 613 = 100 \times P$$
$$\frac{9195}{100} = \frac{100 \times P}{100}$$
$$91.95 = P$$
or about 92 medals

**19.**
$$\frac{54.9}{100} = \frac{P}{15,253}$$
$$54.9 \times 15,253 = 100 \times P$$
$$\frac{837,389.7}{100} = \frac{100 \times P}{100}$$
$$8373.897 = P$$
or about 8374 goals

## Finding What Percent One Number is of Another (page 71)

**1.**
$$\frac{16}{32} = \frac{r}{100}$$
$$16 \times 100 = 32 \times r$$
$$\frac{1600}{32} = \frac{32 \times r}{32}$$
$$50 = r$$

**2.**
$$\frac{12.5}{15} = \frac{r}{100}$$
$$12.5 \times 100 = 15 \times r$$
$$\frac{1250}{15} = \frac{15 \times r}{15}$$
$$83\tfrac{1}{3} \text{ or } 83.\overline{3} = r$$

**3.**
$$\frac{18}{150} = \frac{r}{100}$$
$$18 \times 100 = 150 \times r$$
$$\frac{1800}{150} = \frac{150 \times r}{150}$$
$$12 = r$$

**4.** $B = 50,\ P = 6$
$$\frac{6}{50} = \frac{r}{100}$$
$$6 \times 100 = 50 \times r$$
$$\frac{600}{50} = \frac{50 \times r}{50}$$
$$12 = r$$

**5.** $P = 22.8$, $B = 456$

$$\frac{22.8}{456} = \frac{r}{100}$$
$$22.8 \times 100 = 456 \times r$$
$$\frac{2280}{456} = \frac{456 \times r}{456}$$
$$5 = r$$

**6.** $P = 12$, $B = 32$

$$\frac{12}{32} = \frac{r}{100}$$
$$12 \times 100 = 32 \times r$$
$$\frac{1200}{32} = \frac{32 \times r}{32}$$
$$37.5 = r$$

**7.** $P = 15$, $B = 25$

$$\frac{15}{25} = \frac{r}{100}$$
$$15 \times 100 = 25 \times r$$
$$\frac{1500}{25} = \frac{25 \times r}{25}$$
$$60 = r$$

**8.** $P = 36$, $B = 400$

$$\frac{36}{400} = \frac{r}{100}$$
$$36 \times 100 = 400 \times r$$
$$\frac{3600}{400} = \frac{400 \times r}{400}$$
$$9 = r$$

**9.** $P = 6$, $B = 120$

$$\frac{6}{120} = \frac{r}{100}$$
$$6 \times 100 = 120 \times r$$
$$\frac{600}{120} = \frac{120 \times r}{120}$$
$$5 = r$$

**10.** $P = 16$, $B = 80$

$$\frac{16}{80} = \frac{r}{100}$$
$$16 \times 100 = 80 \times r$$
$$\frac{1600}{80} = \frac{80 \times r}{80}$$
$$20 = r$$

**11.** $P = 108$, $B = 150$

$$\frac{108}{150} = \frac{r}{100}$$
$$108 \times 100 = 150 \times r$$
$$\frac{10,800}{150} = \frac{150 \times r}{150}$$
$$72 = r$$

**12.** $P = 0.85$, $B = 1.7$

$$\frac{0.85}{1.7} = \frac{r}{100}$$
$$0.85 \times 100 = 1.7 \times r$$
$$\frac{85}{1.7} = \frac{1.7 \times r}{1.7}$$
$$50 = r$$

**13.** $P = 5.25$, $B = 21$

$$\frac{5.25}{21} = \frac{r}{100}$$
$$5.25 \times 100 = 21 \times r$$
$$\frac{525}{21} = \frac{21 \times r}{21}$$
$$25 = r$$

**14.** $P = 2$, $B = 16$

$$\frac{2}{16} = \frac{r}{100}$$
$$2 \times 100 = 16 \times r$$
$$\frac{200}{16} = \frac{16 \times r}{16}$$
$$12.5 = r$$

**15.** $P = 60$, $B = 180$

$$\frac{60}{180} = \frac{r}{100}$$
$$60 \times 100 = 180 \times r$$
$$\frac{6000}{180} = \frac{180 \times r}{180}$$
$$33\frac{1}{3} = 33.\overline{3} = r$$

**16.**

$$\frac{675}{900} = \frac{r}{100}$$
$$675 \times 100 = 900 \times r$$
$$\frac{67,500}{900} = \frac{900 \times r}{900}$$
$$75 = r$$

Charles drove 75% of the trip.

**17.**

$$\frac{16}{25} = \frac{r}{100}$$
$$16 \times 100 = 25 \times r$$
$$\frac{1600}{25} = \frac{25 \times r}{25}$$
$$64 = r$$

64% of the bites resulted in a catch.

**18.**

$$\frac{8}{20} = \frac{r}{100}$$
$$8 \times 100 = 20 \times r$$
$$\frac{800}{20} = \frac{20 \times r}{20}$$
$$40 = r$$

40% of the bites resulted in a catch.

**19.**

$$\frac{400}{1600} = \frac{r}{100}$$
$$400 \times 100 = 1600 \times r$$
$$\frac{40,000}{1600} = \frac{1600 \times r}{1600}$$
$$25 = r$$

25% of the total expense was gasoline.

## Finding a Number when a Percent is Given (page 72)

**1.**

$$\frac{9.1}{B} = \frac{175}{100}$$
$$9.1 \times 100 = 175 \times B$$
$$\frac{910}{175} = \frac{175 \times B}{175}$$
$$5\frac{1}{5} \text{ or } 5.2 = B$$

**2.**

$$\frac{300}{B} = \frac{75}{100}$$
$$300 \times 100 = 75 \times B$$
$$\frac{30,000}{75} = \frac{75 \times B}{75}$$
$$400 = B$$

**3.**

$$\frac{72}{B} = \frac{60}{100}$$
$$72 \times 100 = 60 \times B$$
$$\frac{7200}{60} = \frac{60 \times B}{60}$$
$$120 = B$$

**4.**

$$\frac{88}{B} = \frac{80}{100}$$
$$88 \times 100 = 80 \times B$$
$$\frac{8800}{80} = \frac{80 \times B}{80}$$
$$110 = B$$

**5.**

$$\frac{3}{B} = \frac{14\frac{2}{7}}{100}$$
$$3 \times 100 = 14\frac{2}{7} \times B$$
$$\frac{300}{14\frac{2}{7}} = \frac{14\frac{2}{7} \times B}{14\frac{2}{7}}$$
$$300 \div \frac{100}{7} = B$$
$$300 \times \frac{7}{100} = B$$
$$21 = B$$

**6.**

$$\frac{42}{B} = \frac{600}{100}$$
$$42 \times 100 = 600 \times B$$
$$\frac{4200}{600} = \frac{600 \times B}{600}$$
$$7 = B$$

**7.** $P = 56$, $r = 70$

$$\frac{56}{B} = \frac{70}{100}$$
$$56 \times 100 = 70 \times B$$
$$\frac{5600}{70} = \frac{70 \times B}{70}$$
$$80 = B$$

**8.** $P = 3$, $r = 15$

$$\frac{3}{B} = \frac{15}{100}$$
$$3 \times 100 = 15 \times B$$
$$\frac{300}{15} = \frac{15 \times B}{15}$$
$$20 = B$$

**9.** $P = 5.6$, $r = 28$

$$\frac{5.6}{B} = \frac{28}{100}$$
$$5.6 \times 100 = 28 \times B$$
$$\frac{560}{28} = \frac{28 \times B}{28}$$
$$20 = B$$

**10.** $P = 8.5$, $r = 50$

$$\frac{8.5}{B} = \frac{50}{100}$$
$$8.5 \times 100 = 50 \times B$$
$$\frac{850}{50} = \frac{50 \times B}{50}$$
$$17 = B$$

**11.** $P = 6$, $r = 1$

$$\frac{6}{B} = \frac{1}{100}$$
$$6 \times 100 = B \times 1$$
$$600 = B$$

**12.** $P = 20$, $r = 100$

$$\frac{20}{B} = \frac{100}{100}$$
$$20 \times 100 = B \times 100$$
$$\frac{2000}{100} = \frac{B \times 100}{100}$$
$$20 = B$$

**13.** $P = 60$, $r = 30$

$$\frac{60}{B} = \frac{30}{100}$$
$$60 \times 100 = B \times 30$$
$$\frac{6000}{30} = \frac{B \times 30}{30}$$
$$200 = B$$

**14.** $P = 18$, $r = 30$

$$\frac{18}{B} = \frac{30}{100}$$
$$18 \times 100 = 30 \times B$$
$$\frac{1800}{30} = \frac{30 \times B}{30}$$
$$60 = B$$

**15.** $P = 35$, $r = 125$

$$\frac{35}{B} = \frac{125}{100}$$
$$35 \times 100 = 125 \times B$$
$$\frac{3500}{125} = \frac{125 \times B}{125}$$
$$28 = B$$

**16.** $P = 72$, $r = 75$

$$\frac{72}{B} = \frac{75}{100}$$
$$72 \times 100 = 75 \times B$$
$$\frac{7200}{75} = \frac{75 \times B}{75}$$
$$96 = B$$

**17.** $P = 0.891$, $r = 99$

$$\frac{0.891}{B} = \frac{99}{100}$$
$$0.891 \times 100 = 99 \times B$$
$$\frac{89.1}{99} = \frac{99 \times B}{99}$$
$$0.9 = B$$

**18.** $P = 56$, $r = 350$

$$\frac{56}{B} = \frac{350}{100}$$
$$56 \times 100 = B \times 350$$
$$\frac{5600}{350} = \frac{B \times 350}{350}$$
$$16 = B$$

**19.**

$$\frac{5}{100} = \frac{36}{B}$$
$$36 \times 100 = 5 \times B$$
$$\frac{3600}{5} = \frac{5 \times B}{5}$$
shrimp: $720 = B$

**20.**

$$\frac{10}{100} = \frac{\$120,411}{B}$$
$$120,411 \times 100 = 10 \times B$$
$$\frac{12,041,100}{10} = \frac{10B}{10}$$
total winnings: $\$1,204,110 = B$

## Percent, Base, Rate (page 73)

**1.** $r = 5$, $B = 75$
$$\frac{5}{100} = \frac{P}{75}$$
$$100 \times P = 5 \times 75$$
$$P = 3.75$$

**2.** $r = 35$, $B = 400$
$$\frac{35}{100} = \frac{P}{400}$$
$$35 \times 400 = 100 \times P$$
$$140 = P$$

**3.** $B = 80$, $P = 22$
$$\frac{r}{100} = \frac{22}{80}$$
$$80 \times r = 22 \times 100$$
$$r = 27.5$$

**4.** $B = 600$, $P = 75$
$$\frac{r}{100} = \frac{75}{600}$$
$$r \times 600 = 100 \times 75$$
$$r = 12.5$$

**5.** $r = 8$, $B = 800$
$$\frac{8}{100} = \frac{P}{800}$$
$$8 \times 800 = 100 \times P$$
$$64 = P$$

**6.** $r = 55$, $P = 165$
$$\frac{55}{100} = \frac{165}{B}$$
$$55 \times B = 100 \times 165$$
$$B = 300$$

**7.** $r = 70$, $B = 80$
$$\frac{70}{100} = \frac{P}{80}$$
$$70 \times 80 = 100 \times P$$
$$56 = P$$

**8.** $r = 30$, $B = 250$
$$\frac{30}{100} = \frac{P}{250}$$
$$30 \times 250 = 100 \times P$$
$$75 = P$$

**9.** $r = 40$, $P = 200$
$$\frac{40}{100} = \frac{200}{B}$$
$$40 \times B = 200 \times 100$$
$$B = 500$$

**10.** $P = 22$, $B = 220$
$$\frac{22}{220} = \frac{r}{100}$$
$$22 \times 100 = 220 \times r$$
$$10 = r$$

**11.** $P = 40$, $B = 250$
$$\frac{40}{250} = \frac{r}{100}$$
$$40 \times 100 = 250 \times r$$
$$16 = r$$

**12.** $P = 90$, $r = 72$
$$\frac{90}{B} = \frac{72}{100}$$
$$90 \times 100 = 72 \times B$$
$$125 = B$$

**13.** $B = 28$, $P = 14$
$$\frac{r}{100} = \frac{14}{28}$$
$$r \times 28 = 14 \times 100$$
$$r = 50$$

**14.** $r = 5$, $B = 120$
$$\frac{5}{100} = \frac{P}{120}$$
$$5 \times 120 = 100 \times P$$
$$6 = P$$

**15.** $r = 64$, $P = 21$
$$\frac{64}{100} = \frac{21}{B}$$
$$64 \times B = 21 \times 100$$
$$B = 32.8125$$

**16.** $B = 400$, $P = 96$
$$\frac{96}{400} = \frac{r}{100}$$
$$96 \times 100 = 400 \times r$$
$$24 = r$$

**17.** $r = 12.5$, $P = 7$
$$\frac{12.5}{100} = \frac{7}{B}$$
$$12.5 \times B = 7 \times 100$$
$$B = 56$$

**18.** $r = 600$, $P = 474$
$$\frac{600}{100} = \frac{474}{B}$$
$$600 \times B = 100 \times 474$$
$$B = 79$$

**19.** $P = 16$, $r = 20$
$$\frac{16}{B} = \frac{20}{100}$$
$$20 \times B = 16 \times 100$$
$$B = 80$$

**20.** $r = 46.4$, $B = 500$
$$\frac{46.4}{100} = \frac{P}{500}$$
$$46.4 \times 500 = P \times 100$$
$$232 = P$$

**21.** $B = 3600$, $P = 9$
$$\frac{r}{100} = \frac{9}{3600}$$
$$3600 \times r = 9 \times 100$$
$$r = 0.25$$

**22.** $B = 120$, $P = 30$
$$\frac{30}{120} = \frac{r}{100}$$
$$30 \times 100 = 120 \times r$$
$$25 = r$$

**23.** $r = 87.5$, $B = 160$
$$\frac{87.5}{100} = \frac{P}{160}$$
$$87.5 \times 160 = 100 \times P$$
$$140 = P$$

**24.** $r = 22.5$, $P = 31.5$
$$\frac{22.5}{100} = \frac{31.5}{B}$$
$$22.5 \times B = 100 \times 31.5$$
$$B = 140$$

**25.** $r = 5$, $B = 456$
$$\frac{5}{100} = \frac{P}{456}$$
$$5 \times 456 = 100 \times P$$
$$22.8 = P$$

**26.** $P = 794.5$, $r = 350$
$$\frac{794.5}{B} = \frac{350}{100}$$
$$794.5 \times 100 = 350 \times B$$
$$227 = B$$

**27.** $B = 80$, $P = 25$
$$\frac{r}{100} = \frac{25}{80}$$
$$80 \times r = 25 \times 100$$
$$r = 31.25$$

**28.** $r = 1.1$, $B = 2000$
$$\frac{1.1}{100} = \frac{P}{2000}$$
$$1.1 \times 2000 = 100 \times P$$
$$22 = P$$

**29.** $P = 49$, $B = 56$
$$\frac{49}{56} = \frac{r}{100}$$
$$49 \times 100 = 56 \times r$$
$$87.5 = r$$

**30.** $P = 17.85$, $B = 85$
$$\frac{17.85}{85} = \frac{r}{100}$$
$$17.85 \times 100 = r \times 85$$
$$21 = r$$

**31.** $r = 2.5$, $B = 80$
$$\frac{2.5}{100} = \frac{P}{80}$$
$$2.5 \times 80 = 100 \times P$$
$$2 = P$$

**32.** $P = 11.97$, $r = 9$
$$\frac{11.97}{B} = \frac{9}{100}$$
$$11.97 \times 100 = 9 \times B$$
$$133 = B$$

**33.** $r = 460$, $B = 4.5$

$$\frac{460}{100} = \frac{P}{4.5}$$
$$460 \times 4.5 = 100 \times P$$
$$20.7 = P$$

**34.** $r = 4.25$, $B = 600$

$$\frac{4.25}{100} = \frac{P}{600}$$
$$4.25 \times 600 = P \times 100$$
$$25.5 = P$$

**35.** $P = 3$, $r = 14\frac{2}{7}$

$$\frac{3}{B} = \frac{14\frac{2}{7}}{100}$$
$$14\frac{2}{7} \times B = 3 \times 100$$
$$B = 300 \div 14\frac{2}{7}$$
$$B = \frac{300}{1} \times \frac{7}{100}$$
$$B = 21$$

**36.** $P = 48$, $B = 120$

$$\frac{48}{120} = \frac{r}{100}$$
$$48 \times 100 = r \times 120$$
$$40 = r$$

**37.** $P = 30$, $B = 80$
$$\frac{30}{80} = \frac{r}{100}$$
$$30 \times 100 = 80 \times r$$
$$37.5 = r$$

**38.** $P = 43.5$, $r = 7.5$
$$\frac{43.5}{B} = \frac{7.5}{100}$$
$$43.5 \times 100 = 7.5 \times B$$
$$580 = B$$

**39.** $P = 6.38$, $r = 145$
$$\frac{6.38}{B} = \frac{145}{100}$$
$$145 \times B = 6.38 \times 100$$
$$B = 4.4$$

**40.** $P = 511$, $r = 7.3$
$$\frac{511}{B} = \frac{7.3}{100}$$
$$511 \times 100 = B \times 7.3$$
$$7000 = B$$

## Percent of Change (Gain/Loss) (page 74)

**1.** $11   **2.** $8   **3.** $15   **4.** $12   **5.** $40   **6.** $50   **7.** $18   **8.** $4   **9.** $15   **10.** $2

**11.**
$$\begin{array}{r} \$36.80 \\ -\ \$32.00 \\ \hline \$\ 4.80 \end{array}$$

$$\frac{4.80}{32.00} = \frac{r}{100}$$
$$4.80 \times 100 = 32.00 \times r$$
$$480 = 32.00 \times r$$
$$\frac{480}{32.00} = \frac{32.00 \times r}{32.00}$$
$$15 = r$$

**12.**
$$\begin{array}{r} \$4 \\ -\ \$3 \\ \hline \$1 \end{array}$$

$$\frac{1}{4} = \frac{r}{100}$$
$$1 \times 100 = 4 \times r$$
$$100 = 4 \times r$$
$$\frac{100}{4} = \frac{4 \times r}{4}$$
$$25 = r$$

**13.**
$$\begin{array}{r} \$150 \\ -\ \$120 \\ \hline \$\ 30 \end{array}$$

$$\frac{30}{120} = \frac{r}{100}$$
$$30 \times 100 = 120 \times r$$
$$3000 = 120 \times r$$
$$\frac{3000}{120} = \frac{120 \times r}{120}$$
$$25 = r$$

**14.**
$$\begin{array}{r} \$100 \\ -\ \$\ 72 \\ \hline \$\ 28 \end{array}$$

$$\frac{28}{100} = \frac{r}{100}$$
$$28 \times 100 = 100 \times r$$
$$2800 = 100 \times r$$
$$\frac{2800}{100} = \frac{100 \times r}{100}$$
$$28 = r$$

**15.**
$$\begin{array}{r} \$6.00 \\ -\ \$4.20 \\ \hline \$1.80 \end{array}$$

$$\frac{1.80}{6.00} = \frac{r}{100}$$
$$1.80 \times 100 = 6.00 \times r$$
$$180 = 6.00 \times r$$
$$\frac{180}{6.00} = \frac{6.00 \times r}{6.00}$$
$$30 = r$$

**16.**
$$\begin{array}{r} \$250 \\ -\ \$200 \\ \hline \$\ 50 \end{array}$$

$$\frac{50}{250} = \frac{r}{100}$$
$$50 \times 100 = 250 \times r$$
$$5000 = 250 \times r$$
$$\frac{5000}{250} = \frac{250 \times r}{250}$$
$$20 = r$$

**17.**
$$\begin{array}{r} \$49.00 \\ -\ \$43.12 \\ \hline \$\ 5.88 \end{array}$$

$$\frac{5.88}{49} = \frac{r}{100}$$
$$5.88 \times 100 = 49 \times r$$
$$\frac{588}{49} = \frac{49 \times r}{49}$$
$$12 = r$$

**18.**
$$\begin{array}{r} 40 \\ -\ 20 \\ \hline 20 \end{array}$$

$$\frac{20}{20} = \frac{r}{100}$$
$$20 \times 100 = 20 \times r$$
$$2000 = 20 \times r$$
$$\frac{2000}{20} = \frac{20 \times r}{20}$$
$$100 = r$$

**19.**
$$\begin{array}{r} \$1000 \\ -\ \$\ 550 \\ \hline \$\ 450 \end{array}$$

$$\frac{450}{1000} = \frac{r}{100}$$
$$450 \times 100 = 1000 \times r$$
$$\frac{45,000}{1000} = \frac{1000 \times r}{1000}$$
$$45 = r$$

**20.**
$$\begin{array}{r} 150 \\ -\ 120 \\ \hline 30 \end{array}$$

$$\frac{30}{120} = \frac{r}{100}$$
$$30 \times 100 = 120 \times r$$
$$3000 = 120 \times r$$
$$\frac{0000}{120} = \frac{120 \times r}{120}$$
$$25 = r$$

**21.**
$$\begin{array}{r} 35 \\ -\ 7 \\ \hline 28 \end{array}$$

$$\frac{28}{35} = \frac{r}{100}$$
$$28 \times 100 = 35 \times r$$
$$\frac{2800}{35} = \frac{35 \times r}{35}$$
$$80 = r$$

**22.**
$$\begin{array}{r} \$550 \\ -\ \$429 \\ \hline \$121 \end{array}$$

$$\frac{121}{550} = \frac{r}{100}$$
$$121 \times 100 = 550 \times r$$
$$12,100 = 550 \times r$$
$$\frac{12,100}{550} = \frac{550 \times r}{550}$$
$$22 = r$$

## Discount (pages 75–76)

**1.**
$$\begin{array}{r} \$45.00 \\ -\ \$\ 9.00 \\ \hline \$36.00 \end{array}$$

**2.**
$$\begin{array}{r} \$26.20 \\ -\ \$\ 3.54 \\ \hline \$22.66 \end{array}$$

**3.**
$$\begin{array}{r} \$18.99 \\ -\ \$\ 5.60 \\ \hline \$13.39 \end{array}$$

**4.**
$$\begin{array}{r} \$285.40 \\ -\ \$\ 29.80 \\ \hline \$255.60 \end{array}$$

**5.**
$$\begin{array}{r} \$29.75 \\ -\ \$9.25 \\ \hline \$20.50 \end{array}$$

**6.**
$$\begin{array}{r} \$66.90 \\ -\ \$\ 8.50 \\ \hline \$58.40 \end{array}$$

**7.**
$$\begin{array}{r} \$305 \\ \times\ 0.20 \\ \hline \$61 \end{array}$$

**8.**
$$\begin{array}{r} \$118.98 \\ \times\ 0.333 \\ \hline 35694 \\ 3\ 5694 \\ 35\ 694 \\ \hline \$39.62034 \end{array}$$ or \$39.62

**9.**
$$\begin{array}{r} \$51.92 \\ \times\ 0.125 \\ \hline 25960 \\ 1\ 0384 \\ 5\ 192 \\ \hline \$6.49000 \end{array}$$ or \$6.49

**10.**
$$\begin{array}{r} \$29.40 \\ \times\ 0.05 \\ \hline \$1.4700 \end{array}$$ or \$1.47

**11.**
$$\begin{array}{r} \$209.90 \\ \times\ 0.25 \\ \hline 10\ 4950 \\ 41\ 980 \\ \hline \$52.4750 \end{array}$$ or \$52.48

**12.**
$$\begin{array}{r} \$22.80 \\ \times\ 0.10 \\ \hline \$2.280 \end{array}$$ or \$2.28

**13.**
$$\begin{array}{r} \$100.95 \\ \times\ 0.125 \\ \hline 50475 \\ 2\ 0190 \\ 10\ 095 \\ \hline \$12.61875 \end{array}$$ or \$12.62
$$\begin{array}{r} \$100.95 \\ -\ \$\ 12.62 \\ \hline \$\ 88.33 \end{array}$$

**14.**
$$\begin{array}{r} \$316.00 \\ \times\ 0.20 \\ \hline \$63.2000 \end{array}$$ or \$63.20
$$\begin{array}{r} \$316.00 \\ -\ \$\ 63.20 \\ \hline \$252.80 \end{array}$$

**15.**
$$\begin{array}{r} \$15.90 \\ \times\ 0.12 \\ \hline 3180 \\ 1\ 590 \\ \hline \$1.9080 \end{array}$$ or \$1.91
$$\begin{array}{r} \$15.90 \\ -\ \$\ 1.91 \\ \hline \$13.99 \end{array}$$

**16.**
$$\begin{array}{r} \$17.00 \\ 0.11 \\ \hline 1700 \\ 1\ 700 \\ \hline \$1.8700 \end{array}$$ or \$1.87
$$\begin{array}{r} \$17.00 \\ -\ \$\ 1.87 \\ \hline \$15.13 \end{array}$$

**17.**
$$\begin{array}{r} \$229.99 \\ \times\ 0.333 \\ \hline 68997 \\ 6\ 8997 \\ 68\ 997 \\ \hline \$76.58667 \end{array}$$ or \$76.59
$$\begin{array}{r} \$229.99 \\ -\ \$\ 76.59 \\ \hline \$153.40 \end{array}$$

**18.**
$$\begin{array}{r} \$69.95 \\ \times\ 0.10 \\ \hline \$6.9950 \end{array}$$ or \$7.00
$$\begin{array}{r} \$69.95 \\ -\ \$7.00 \\ \hline \$62.95 \end{array}$$

**19.**
$$\begin{array}{r} \$110.00 \\ -\ \$55.00 \\ \hline \$\ 55.00 \end{array}$$

**20.**
$$\begin{array}{r} \$3.72 \\ -\ \$1.50 \\ \hline \$2.22 \end{array}$$

**21.**
$$\begin{array}{r} \$75.78 \\ \times\ 0.166 \\ \hline 45468 \\ 4\ 5468 \\ 7\ 578 \\ \hline \$12.57948 \end{array} \text{ or } \$12.58$$
$$\begin{array}{r} \$75.78 \\ -\ \$12.58 \\ \hline \$63.20 \end{array}$$

**22.**
$$\begin{array}{r} \$83.79 \\ \times\ 0.35 \\ \hline 4\ 1895 \\ 25\ 137 \\ \hline \$29.3265 \end{array} \text{ or } \$29.33$$
$$\begin{array}{r} \$83.79 \\ -\ \$29.33 \\ \hline \$54.46 \end{array}$$

**23.**
$$\begin{array}{r} \$19.80 \\ -\ \$\ 4.95 \\ \hline \$14.85 \end{array}$$

**24.**
$$\begin{array}{r} \$655 \\ \times\ 0.20 \\ \hline \$131.00 \end{array}$$
$$\begin{array}{r} \$655 \\ -\ \$131 \\ \hline \$524 \end{array}$$

**25.**
$$\begin{array}{r} \$26.00 \\ \times\ \ 0.40 \\ \hline \$10.4000 \end{array}$$
$$\begin{array}{r} \$26.00 \\ -\ \$10.40 \\ \hline \$15.60 \end{array}$$

**26.**
$$\begin{array}{r} \$79.89 \\ -\ \$16.99 \\ \hline \$62.90 \end{array}$$

**27.**
$$\begin{array}{r} \$31 \\ \times\ 0.05 \\ \hline \$1.55 \end{array}$$
$$\begin{array}{r} \$31.00 \\ -\ \$\ 1.55 \\ \hline \$29.45 \end{array}$$

**28.**
$$\begin{array}{r} \$559.95 \\ -\ \$\ 17.76 \\ \hline \$542.19 \end{array}$$

**29.**
$$\begin{array}{r} \$210 \\ \times\ 0.10 \\ \hline \$21.00 \end{array}$$
$$\begin{array}{r} \$210 \\ -\ \$21 \\ \hline \$189 \end{array}$$

**30.**
$$\begin{array}{r} \$407.50 \\ -\ \$\ 65.20 \\ \hline \$342.30 \end{array}$$

**31.** $100\% - 13\% = 87\%$
$$\frac{P}{1{,}246} = \frac{87}{100}$$
$$P \times 100 = 1{,}246 \times 87$$
$$P \times 100 = 108{,}402$$
$$\frac{P \times 100}{100} = \frac{108{,}402}{100}$$
$$P = \$1084.02$$

**32.** $100\% - 22\% = 78\%$
$$\frac{P}{327} = \frac{78}{100}$$
$$P \times 100 = 327 \times 78$$
$$P \times 100 = 25{,}506$$
$$\frac{P \times 100}{100} = \frac{25{,}506}{100}$$
$$P = \$255.06$$

**33.** $100\% - 15\% = 85\%$
$$\frac{P}{126} = \frac{85}{100}$$
$$P \times 100 = 126 \times 85$$
$$P \times 100 = 10{,}710$$
$$\frac{P \times 100}{100} = \frac{10{,}710}{100}$$
$$P = \$107.10$$

**34.** $100\% - 4\% = 96\%$
$$\frac{P}{46} = \frac{96}{100}$$
$$P \times 100 = 46 \times 96$$
$$P \times 100 = 4416$$
$$\frac{P \times 100}{100} = \frac{4416}{100}$$
$$P = \$44.16$$

**35.** $100\% - 28\% = 72\%$
$$\frac{P}{59.85} = \frac{72}{100}$$
$$P \times 100 = 59.85 \times 72$$
$$P \times 100 = 4309.20$$
$$\frac{P \times 100}{100} = \frac{4309.20}{100}$$
$$P = \$43.09$$

**36.** $100\% - 18\% = 82\%$
$$\frac{P}{4560} = \frac{82}{100}$$
$$P \times 100 = 4560 \times 82$$
$$P \times 100 = 373{,}920$$
$$\frac{P \times 100}{100} = \frac{373{,}920}{100}$$
$$P = \$3739.20$$

**37.**
$$\frac{6.40}{16} = \frac{r}{100}$$
$$6.40 \times 100 = r \times 16$$
$$640 = r \times 16$$
$$\frac{640}{16} = \frac{r \times 16}{16}$$
$$40 = r \text{ discount rate}$$

**38.**
$$\frac{P}{6.80} = \frac{15}{100}$$
$$P \times 100 = 6.80 \times 15$$
$$P \times 100 = 102$$
$$\frac{P \times 100}{100} = \frac{102}{100}$$
$$P = \$1.02 \text{ discount}$$

**39.**
$$\frac{7.98}{19.95} = \frac{r}{100}$$
$$7.98 \times 100 = 19.95 \times r$$
$$798 = 19.95 \times r$$
$$\frac{798}{19.95} = \frac{19.95 \times r}{19.95}$$
$$40 = r \text{ discount rate}$$

**40.**
$$\frac{P}{12} = \frac{35}{100}$$
$$P \times 100 = 12 \times 35$$
$$P \times 100 = 420$$
$$\frac{P \times 100}{100} = \frac{420}{100}$$
$$P = \$4.20 \text{ discount}$$

**41.**
$$\frac{20}{80} = \frac{r}{100}$$
$$20 \times 100 = 80 \times r$$
$$2000 = 80 \times r$$
$$\frac{2000}{80} = \frac{80 \times r}{80}$$
$$25 = r \text{ discount rate}$$

**42.**
$$\frac{8.50}{B} = \frac{20}{100}$$
$$8.50 \times 100 = B \times 20$$
$$\frac{850}{20} = \frac{B \times 20}{20}$$
$$\$42.50 = B \text{ original price}$$

**43.**
$$\frac{29.40}{B} = \frac{60}{100}$$
$$29.40 \times 100 = B \times 60$$
$$\frac{2940}{60} = \frac{B \times 60}{60}$$
$$\$49 = B \text{ original price}$$

**44.**
$$\frac{P}{20} = \frac{10}{100}$$
$$P \times 100 = 10 \times 20$$
$$P \times 100 = 200$$
$$\frac{P \times 100}{100} = \frac{200}{100}$$
$$P = \$2 \text{ discount}$$

**45.**
$$\frac{9.75}{B} = \frac{25}{100}$$
$$9.75 \times 100 = B \times 25$$
$$975 = B \times 25$$
$$\frac{975}{25} = \frac{B \times 25}{25}$$
$$\$39 = B \text{ original price}$$

**46.**
$$\frac{P}{470} = \frac{20}{100}$$
$$P \times 100 = 470 \times 20$$
$$P \times 100 = 9400$$
$$\frac{P \times 100}{100} = \frac{9400}{100}$$
$$P = \$94 \text{ discount}$$

**47.**
$$\frac{P}{693.95} = \frac{25}{100}$$
$$P \times 100 = 693.95 \times 25$$
$$P \times 100 = 17{,}348.75$$
$$\frac{P \times 100}{100} = \frac{17{,}348.75}{100}$$
$$P = \$173.4875 \text{ or } \$173.49$$
$$\begin{array}{r} \$693.95 \\ -\ \$173.49 \\ \hline \$520.46 \text{ sale price} \end{array}$$

**48.**
$$\frac{34.59}{230.60} = \frac{r}{100}$$
$$34.59 \times 100 = 230.60 \times r$$
$$3459 = 230.60 \times r$$
$$\frac{3459}{230.60} = \frac{230.60 \times r}{230.60}$$
$$15 = r \text{ discount rate}$$

**49.**
$$\frac{8.25}{B} = \frac{16.6}{100}$$
$$8.25 \times 100 = B \times 16.6$$
$$\frac{825}{16.6} = \frac{B \times 16.6}{16.6}$$
$$49.698795 \text{ or } \$49.70 \text{ original price}$$

**50.**
$$\frac{P}{72.98} = \frac{10}{100}$$
$$P \times 100 = 72.98 \times 10$$
$$P \times 100 = 729.8$$
$$\frac{P \times 100}{100} = \frac{729.8}{100}$$
$$P = \$7.298 \text{ or } \$7.30 \text{ discount}$$
$$\begin{array}{r} \$72.98 \\ -\ \$7.30 \\ \hline \$65.68 \text{ sale price} \end{array}$$

**51.**
$$\frac{P}{579} = \frac{15}{100}$$
$$P \times 100 = 579 \times 15$$
$$P \times 100 = 8685$$
$$\frac{P \times 100}{100} = \frac{8685}{100}$$
$$P = \$86.85 \text{ discount}$$

## Profit (Markup) (pages 77–78)

**1.**
$$\begin{array}{r} \$72,300 \\ \times\ 0.25 \\ \hline \$18,075 \end{array} \qquad \begin{array}{r} \$72,300 \\ +\ \$18,075 \\ \hline \$90,375 \end{array}$$

**2.**
$$\begin{array}{r} \$38,000 \\ \times\ 0.35 \\ \hline \$13,300 \end{array} \qquad \begin{array}{r} \$38,000 \\ +\ \$13,300 \\ \hline \$51,300 \end{array}$$

**3.**
$$\begin{array}{r} \$26,500 \\ \times\ 0.20 \\ \hline \$5,300 \end{array} \qquad \begin{array}{r} \$26,500 \\ +\ \$\ 5,300 \\ \hline \$31,800 \end{array}$$

**4.**
$$\begin{array}{r} \$30,950 \\ \times\ 0.10 \\ \hline \$3095 \end{array} \qquad \begin{array}{r} \$30,950 \\ +\$3095 \\ \hline \$34,045 \end{array}$$

**5.**
$$\begin{array}{r} \$27,000 \\ \times\ 0.15 \\ \hline \$4050 \end{array} \qquad \begin{array}{r} \$27,000 \\ +\ \$\ 4050 \\ \hline \$31,050 \end{array}$$

**6.**
$$\begin{array}{r} \$40,000 \\ \times\ 0.18 \\ \hline \$7200 \end{array} \qquad \begin{array}{r} \$40,000 \\ +\ \$\ 7200 \\ \hline \$47,200 \end{array}$$

**7.**
$$\frac{P}{115} = \frac{100}{100}$$
$$P \times 100 = 115 \times 100$$
$$P \times 100 = 11,500$$
$$\frac{P \times 100}{100} = \frac{11,500}{100}$$
$$P = 115$$

**8.**
$$\frac{P}{60} = \frac{25}{100}$$
$$P \times 100 = 60 \times 25$$
$$P \times 100 = 1500$$
$$\frac{P \times 100}{100} = \frac{1500}{100}$$
$$P = 15$$

**9.**
$$\frac{P}{90} = \frac{70}{100}$$
$$P \times 100 = 90 \times 70$$
$$P \times 100 = 6300$$
$$\frac{P \times 100}{100} = \frac{6300}{100}$$
$$P = 63$$

**10.**
$$\frac{P}{600} = \frac{75}{100}$$
$$P \times 100 = 600 \times 75$$
$$P \times 100 = 45,000$$
$$\frac{P \times 100}{100} = \frac{45,000}{100}$$
$$P = 450$$

**11.**
$$\frac{P}{35} = \frac{40}{100}$$
$$P \times 100 = 40 \times 35$$
$$P \times 100 = 1400$$
$$\frac{P \times 100}{100} = \frac{1400}{100}$$
$$P = 14$$

**12.**
$$\frac{P}{500} = \frac{42}{100}$$
$$P \times 100 = 42 \times 500$$
$$P \times 100 = 21,000$$
$$\frac{P \times 100}{100} = \frac{21,000}{100}$$
$$P = 210$$

**13.**
$$\begin{array}{r} \$45.60 \\ \times\ 0.15 \\ \hline 2\ 2800 \\ 4\ 560 \\ \hline \$6.8400 \end{array} \ \text{markup} \qquad \begin{array}{r} \$45.60 \\ +\ \$\ 6.84 \\ \hline \$52.44 \end{array} \text{selling price}$$

**14.**
$$\begin{array}{r} \$27.50 \\ \times\ 0.35 \\ \hline 1\ 3750 \\ 8\ 250 \\ \hline \$9.6250 \end{array} \text{ or } \$9.63 \text{ store income}$$

**15.**
$$\begin{array}{r} \$700 \\ \times\ 0.08 \\ \hline \$56.00 \end{array} \text{ earnings}$$

**16.**
$$\begin{array}{r} \$62.72 \\ -\ \$56.00 \\ \hline \$\ 6.72 \end{array} \begin{array}{l} \text{amount of} \\ \text{markup} \end{array}$$
$$\frac{6.72}{56} = \frac{r}{100}$$
$$56 \times r = 6.72 \times 100$$
$$\frac{56 \times r}{56} = \frac{672}{56}$$
$$r = 12 \text{ percent of markup}$$

**17.**
$$\begin{array}{r} \$49,965 \\ -\ \$32,099 \\ \hline \$17,866 \end{array}$$

**10.**
$$\begin{array}{r} \$139.95 \\ \times\ 0.30 \\ \hline \$41.9850 \end{array} \qquad \begin{array}{r} \$130.95 \\ +\ \$\ 41.99 \\ \hline \$181.94 \end{array} \begin{array}{l} \text{selling price} \\ \text{or }\$41.99 \\ \text{markup} \end{array}$$

## Depreciation (page 79)

1. 35% + 17% + 13% = 65%, 100% − 65% = 35%; 9360 × 0.35 = $3,276.00 3rd year
   35% + 17% + 13% + 11% = 76%, 100% − 76% = 24%; $9360 × 0.24 = $2246.40 4th year
   3rd year − 4th year = value lost: $3276 − $2246.40 = $1029.60 value lost

2. 35% + 17% + 13% + 11% = 76%, 100% − 76% = 24%; $12,632 × 0.24 = $3031.68

3. 35% + 17% = 52%, 100% − 52% = 48%; $21,780 × 0.48 = $10,454.40 2nd year value
   35% + 17% + 13% = 65%, 100% − 65% = 35%; $21,780 × 0.35 = $7623 3rd year value
   $10,454.40 − $7623 = $2831.40 3rd year depreciation

4. 35% + 17% + 13% + 11% = 76%, 100% − 76% = 24%; $8109 × 0.24 = $1946.16 4th year
   35% + 17% + 13% + 11% + 8% = 84%, 100% − 84% = 16%; $8109 × 0.16 = $1297.44 5th year
   4th year value − 5th year value = value lost: $1946.16 − $1297.44 = $648.72 value lost

5. 35% + 17% = 52%, 100% − 52% = 48%; 5872 × 0.48 = $2818.56 2nd year value
   35% + 17% + 13% = 65%, 100% − 65% = 35%; 5872 × 0.35 = $2055.20 3rd year value
   2nd year value − 3rd year value = value lost: $2818.56 − $2055.20 = $763.36 value lost

6. 35% + 17% = 52%, 100% − 52% = 48%; $16,590 × 0.48 = $7963.20

7. $11,723 × 0.16 = $1875.68 5th year value
   $11,723 − $1875.68 = $9847.32 is depreciation over 5 years

8. 1st year percent: 100% − 35% = 65%; first year cash value: $6954 × 0.65 = $4520.10;
   Two years percent: 35% + 17% = 52%, 100% − 52% = 48%; 2nd year cash value: $6954 × 0.48 = 3337.92;
   Three years percent: 35% + 17% + 13% = 65%, 100% − 65% = 35%; 3rd year cash value: $6954 × 0.35 = $2433.90;
   Four years percent: 35% + 17% + 13% + 11% = 76%, 100% − 76% = 24%;
       4th year cash value: $6954 × 0.24 = $1668.96;
   Five years percent: 35% + 17% + 13% + 11% + 8% = 84%, 100% − 84% = 16%;
       5th year cash value: $6954 × 0.16 = $1112.64

## Simple Interest (page 80)

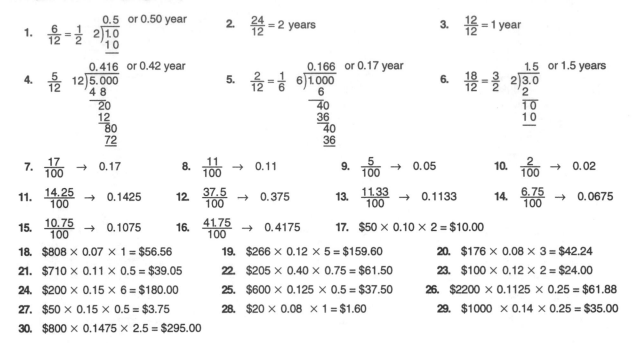

1. $\frac{6}{12} = \frac{1}{2}$  $2\overline{)1.0}$ $\frac{0.5}{}$  or 0.50 year
   $\phantom{2\overline{)}}\underline{1\ 0}$

2. $\frac{24}{12} = 2$ years

3. $\frac{12}{12} = 1$ year

4. $\frac{5}{12}$  $12\overline{)5.000}$ $\frac{0.416}{}$  or 0.42 year
   $\phantom{12\overline{)}}\underline{4\ 8}$
   $\phantom{12\overline{)}}\ \ 20$
   $\phantom{12\overline{)}}\ \ \underline{12}$
   $\phantom{12\overline{)}}\ \ \ 80$
   $\phantom{12\overline{)}}\ \ \ \underline{72}$

5. $\frac{2}{12} = \frac{1}{6}$  $6\overline{)1.000}$ $\frac{0.166}{}$  or 0.17 year
   $\phantom{6\overline{)}}\underline{6}$
   $\phantom{6\overline{)}}\ 40$
   $\phantom{6\overline{)}}\ \underline{36}$
   $\phantom{6\overline{)}}\ 40$
   $\phantom{6\overline{)}}\ \underline{36}$

6. $\frac{18}{12} = \frac{3}{2}$  $2\overline{)3.0}$ $\frac{1.5}{}$  or 1.5 years
   $\phantom{2\overline{)}}\underline{2}$
   $\phantom{2\overline{)}}\overline{1\,0}$
   $\phantom{2\overline{)}}\underline{1\,0}$

7. $\frac{17}{100} \rightarrow 0.17$

8. $\frac{11}{100} \rightarrow 0.11$

9. $\frac{5}{100} \rightarrow 0.05$

10. $\frac{2}{100} \rightarrow 0.02$

11. $\frac{14.25}{100} \rightarrow 0.1425$

12. $\frac{37.5}{100} \rightarrow 0.375$

13. $\frac{11.33}{100} \rightarrow 0.1133$

14. $\frac{6.75}{100} \rightarrow 0.0675$

15. $\frac{10.75}{100} \rightarrow 0.1075$

16. $\frac{41.75}{100} \rightarrow 0.4175$

17. $\$50 \times 0.10 \times 2 = \$10.00$

18. $\$808 \times 0.07 \times 1 = \$56.56$

19. $\$266 \times 0.12 \times 5 = \$159.60$

20. $\$176 \times 0.08 \times 3 = \$42.24$

21. $\$710 \times 0.11 \times 0.5 = \$39.05$

22. $\$205 \times 0.40 \times 0.75 = \$61.50$

23. $\$100 \times 0.12 \times 2 = \$24.00$

24. $\$200 \times 0.15 \times 6 = \$180.00$

25. $\$600 \times 0.125 \times 0.5 = \$37.50$

26. $\$2200 \times 0.1125 \times 0.25 = \$61.88$

27. $\$50 \times 0.15 \times 0.5 = \$3.75$

28. $\$20 \times 0.08 \times 1 = \$1.60$

29. $\$1000 \times 0.14 \times 0.25 = \$35.00$

30. $\$800 \times 0.1475 \times 2.5 = \$295.00$

## Taxes (page 81–82)

1. $\frac{x}{99,000} = \frac{30}{100}$
   $100x = 30 \times 99,000$
   $100x = 2,970,000$
   $\frac{100x}{100} = \frac{2,970,000}{100}$
   $x = \$29,700$

2. $\frac{x}{175,000} = \frac{30}{100}$
   $100x = 30 \times 175,000$
   $100x = 5,250,000$
   $\frac{100x}{100} = \frac{5,250,000}{100}$
   $x = \$52,500$

3. $\frac{x}{50,900} = \frac{30}{100}$
   $100x = 30 \times 50,900$
   $\frac{100x}{100} = \frac{1,527,000}{100}$
   $x = \$15,270$

4. $\frac{x}{78,800} = \frac{30}{100}$
   $100x = 30 \times 78,800$
   $\frac{100x}{100} = \frac{2,364,000}{100}$
   $x = \$23,640$

5. $\frac{x}{225,000} = \frac{30}{100}$
   $100x = 30 \times 225,000$
   $\frac{100x}{100} = \frac{6,750,000}{100}$
   $x = \$67,500$

6. $\frac{x}{460,000} = \frac{30}{100}$
   $100x = 30 \times 460,000$
   $\frac{100x}{100} = \frac{13,800,000}{100}$
   $x = \$138,000$

7. $\frac{x}{88,200} = \frac{30}{100}$
   $100x = 30 \times 88,200$
   $\frac{100x}{100} = \frac{2,646,000}{100}$
   $x = \$26,460$

8. $\frac{x}{97,200} = \frac{30}{100}$
   $100x = 30 \times 97,200$
   $\frac{100x}{100} = \frac{2,916,000}{100}$
   $x = \$29,160$

9. $\frac{9500}{1000} = 9.50$

10. $\frac{54,400}{1000} = 54.40$

11. $\frac{\$9250}{1000} = 9.25$

12. $\frac{\$22,440}{1000} = 22.44$

13. $\frac{70,000}{1000} = 70.00$

14. $\frac{\$107,000}{1000} = 107.00$

15. $\frac{\$84,460}{1000} = \$84.46$

16. $\frac{\$99,970}{1000} = 99.97$

17. $8.5 \times 56.82 = \$482.97$

18. $7.7 \times 56.82 = \$437.51$

19. $29.5 \times 56.82 = \$1676.19$

20. $39.4 \times 56.82 = \$2238.71$

21. $33.7 \times 56.82 = \$1914.83$

22. $\frac{x}{79,000} = \frac{25}{100}$
    $100x = 1,975,000$
    $\frac{100x}{100} = \frac{1,975,000}{100}$
    $x = 19,750$
    $\frac{19,750}{1000} = 19.75$
    19.75

23. $\frac{x}{59,900} = \frac{25}{100}$
    $100x = 1,497,500$
    $\frac{100x}{100} = \frac{1,497,500}{100}$
    $x = 14,975$
    $\frac{14,975}{1000} = 14.975$
    14.975

**24.**
$$\frac{x}{95,000} = \frac{25}{100}$$
$$100x = 2,375,000$$
$$\frac{100x}{100} = \frac{2,375,000}{100}$$
$$x = 23,750$$
$$\frac{23,750}{1000} = 23.75$$

23.75

**25.**
$$\frac{x}{45,000} = \frac{25}{100}$$
$$100x = 1,125,000$$
$$\frac{100x}{100} = \frac{1,125,000}{100}$$
$$x = 11,250$$
$$\frac{11,250}{1000} = 11.25$$

11.25

**26.**
$$\frac{x}{80,000} = \frac{25}{100}$$
$$100x = 2,000,000$$
$$\frac{100x}{100} = \frac{2,000,000}{100}$$
$$x = 20,000$$
$$\frac{20,000}{1000} = 20$$

20

**27.**
$$\frac{x}{72,500} = \frac{25}{100}$$
$$100x = 1,812,500$$
$$\frac{100x}{100} = \frac{1,812,500}{100}$$
$$x = 18,125$$
$$\frac{18,125}{1000} = 18.125$$

18.125

**28.**
$$\frac{x}{54,000} = \frac{25}{100}$$
$$100x = 1,350,000$$
$$\frac{100x}{100} = \frac{1,350,000}{100}$$
$$x = 13,500$$
$$\frac{13,500}{1000} = 13.50$$

13.50

**29.**
$$\frac{x}{30,000} = \frac{25}{100}$$
$$100x = 750,000$$
$$\frac{100x}{100} = \frac{750,000}{100}$$
$$x = 7500$$
$$\frac{7500}{1000} = 7.5$$

7.5

**30.**
$$\frac{x}{63,600} = \frac{25}{100}$$
$$100x = 1,587,500$$
$$\frac{100x}{100} = \frac{1,587,500}{100}$$
$$x = 15,875$$
$$\frac{15,875}{1000} = 15.875$$

15.875

**31.**
$$\frac{x}{04,000} = \frac{25}{100}$$
$$100x = 872,500$$
$$\frac{100x}{100} = \frac{872,500}{100}$$
$$x = 8,725$$
$$\frac{8725}{1000} = 8.725$$

8.725

**32.**
$$\frac{x}{68,500} = \frac{35}{100}$$
$$100x = 2,397,500$$
$$\frac{100x}{100} = \frac{2,397,500}{100}$$
$$x = 23,975$$
$$\frac{23,975}{1000} = 23.975$$

23.975

**33.**
$$\frac{x}{54,000} = \frac{30}{100}$$
$$100x = 1,620,000$$
$$\frac{100x}{100} = \frac{1,620,000}{100}$$
$$x = 16,200$$
$$\frac{16,200}{1000} = 16.20$$

16.20

**34.**
$$\frac{P}{58} = \frac{4.5}{100}$$
$$P \times 100 = 58 \times 4.5$$
$$P \times 100 = 261$$
$$\frac{P \times 100}{100} = \frac{261}{100}$$
$$P = 2.61$$
$$\$58 + \$2.61 = \$60.61$$

**35.**
$$\frac{P}{265} = \frac{5}{100}$$
$$P \times 100 = 1325$$
$$\frac{P \times 100}{100} = \frac{1325}{100}$$
$$P = 13.25$$
$$\$265 + \$13.25 = \$278.25$$

**36.**
$$\frac{P}{18.50} = \frac{6.25}{100}$$
$$P \times 100 = 18.50 \times 6.25$$
$$\frac{P \times 100}{100} = \frac{115.625}{100}$$
$$P = 1.16$$
$$\$18.50 + \$1.16 = \$19.66$$

**37.**
$$\frac{P}{80} = \frac{10}{100}$$
$$P \times 100 = 80 \times 10$$
$$\frac{P \times 100}{100} = \frac{800}{100}$$
$$P = 8$$
$$\$80 + \$8 = \$88$$

**38.**
$$\frac{P}{659.60} = \frac{5.25}{100}$$
$$P \times 100 = 659.60 \times 5.25$$
$$\frac{P \times 100}{100} = \frac{3462.90}{100}$$
$$P = 34.63$$
$$\$659.60 + \$34.63 = \$694.23$$

**39.**
$$\frac{P}{6.59} = \frac{9}{100}$$
$$P \times 100 = 6.59 \times 9$$
$$\frac{P \times 100}{100} = \frac{59.31}{100}$$
$$P = 0.59$$
$$\$6.59 + \$0.59 = \$7.18$$

**40.**
$$\frac{P}{124} = \frac{8.75}{100}$$
$$P \times 100 = 124 \times 8.75$$
$$\frac{P \times 100}{100} = \frac{1085}{100}$$
$$P = 10.85$$
$$\$124 + \$10.85 = \$134.85$$

**41.**
$$\frac{P}{33} = \frac{7}{100}$$
$$P \times 100 = 231$$
$$\frac{P \times 100}{100} = \frac{231}{100}$$
$$P = 2.31$$
$$\$33 + \$2.31 = \$35.31$$

**42.**
$$\frac{P}{20} = \frac{6}{100}$$
$$P \times 100 = 20 \times 6$$
$$P \times 100 = 120$$
$$\frac{P \times 100}{100} = \frac{120}{100}$$
$$P = 1.20$$
$$\$20 + 1.20 = \$21.20$$

**43.** rate    **44.** base    **45.** percentage

**46.**
$$\frac{18}{300} = \frac{r}{100}$$
$$18 \times 100 = 300 \times r$$
$$1800 = 300 \times r$$
$$\frac{1800}{300} = \frac{300 \times r}{300}$$
$$6 = r$$

**47.**
$$\frac{4.41}{B} = \frac{6.25}{100}$$
$$4.41 \times 100 = 6.25 \times B$$
$$441 = 6.25 \times B$$
$$\frac{441}{6.25} = \frac{6.25 \times B}{6.25}$$
$$\$70.56 = B$$

**48.**
$$\frac{13.25}{212} = \frac{r}{100}$$
$$212 \times r = 13.25 \times 100$$
$$\frac{212 \times r}{212} = \frac{1325}{212}$$
$$r = 6.25 \text{ or } 6\frac{1}{4}$$

**49.**
$$\frac{1.50}{B} = \frac{7.5}{100}$$
$$1.50 \times 100 = 7.5 \times B$$
$$150 = 7.5 \times B$$
$$\frac{150}{7.5} = \frac{7.5 \times B}{7.5}$$
$$\$20 = B$$

**50.**
$$\frac{P}{69.98} = \frac{5.75}{100}$$
$$P \times 100 = 69.98 \times 5.75$$
$$\frac{P \times 100}{100} = \frac{402.385}{100}$$
$$P = \$4.02$$

**51.**
$$\frac{2.02}{26} = \frac{r}{100}$$
$$2.02 \times 100 = 26 \times r$$
$$202 = 26 \times r$$
$$\frac{202}{26} = \frac{26 \times r}{26}$$
$$7.7692308 \text{ or } 7.77 = r$$

**52.**
$$\frac{P}{90} = \frac{8}{100}$$
$$P \times 100 = 90 \times 8$$
$$\frac{P \times 100}{100} = \frac{720}{100}$$
$$P = 7.20$$

**53.**
$$\frac{9.81}{B} = \frac{4.5}{100}$$
$$9.81 \times 100 = 4.5 \times B$$
$$\frac{981}{4.5} = \frac{4.5 \times B}{4.5}$$
$$\$218 = B$$

**54.**
$$\frac{P}{221.40} = \frac{4.6}{100}$$
$$P \times 100 = 221.40 \times 4.6$$
$$\frac{P \times 100}{100} = \frac{1018.44}{100}$$
$$P = \$10.18$$

**55.**
$$\frac{P}{139.59} = \frac{5}{100}$$
$$P \times 100 = 5 \times 139.59$$
$$\frac{P \times 100}{100} = \frac{697.95}{100}$$
$$P = \$6.98 \text{ sales tax}$$

**56.**
$$\frac{P}{15.95} = \frac{4.6}{100}$$
$$P \times 100 = 15.95 \times 4.6$$
$$P \times 100 = 73.37$$
$$\frac{P \times 100}{100} = \frac{73.37}{100}$$
$$P = \$0.73 \text{ sales tax}$$

**57.**
$$\frac{22.61}{266} = \frac{r}{100}$$
$$266 \times r = 22.61 \times 100$$
$$\frac{266 \times r}{266} = \frac{2261}{266}$$
$$r = 8.5 \text{ or } 8\frac{1}{2}$$
tax rate

**58.**
$$\frac{0.89}{B} = \frac{5}{100}$$
$$0.89 \times 100 = 5 \times B$$
$$\frac{89}{5} = \frac{5 \times B}{5}$$
$$\$17.80 = B$$
price of the clock